农科英语
文献阅读

Agricultural English Literature Reading

◎张 新 杨 倩 张志勇 主编

中国农业科学技术出版社

图书在版编目(CIP)数据

农科英语文献阅读/张新,杨倩,张志勇主编.--北京:中国农业科学技术出版社,2024.5
ISBN 978-7-5116-6786-1

Ⅰ.①农…　Ⅱ.①张…②杨…③张…　Ⅲ.①农业科学-英语　Ⅳ.①S

中国国家版本馆 CIP 数据核字(2024)第 079083 号

责任编辑　周伟平　崔改泵
责任校对　李向荣
责任印制　姜义伟　王思文

出 版 者	中国农业科学技术出版社
	北京市中关村南大街 12 号　邮编:100081
电　　话	(010) 82106638 (编辑室)　　(010) 82106624 (发行部)
	(010) 82109709 (读者服务部)
网　　址	https://castp.caas.cn
经 销 者	各地新华书店
印 刷 者	北京建宏印刷有限公司
开　　本	185 mm×260 mm　1/16
印　　张	15.5
字　　数	302 千字
版　　次	2024 年 5 月第 1 版　2024 年 5 月第 1 次印刷
定　　价	60.00 元

◆━━ 版权所有・翻印必究 ━━◆

《农科英语文献阅读》编写委员会

主　　编：张　新　杨　倩　张志勇
副主编：杜雄明　李会民　陈双臣
　　　　孔祥军　王　森　李增强
　　　　杨惠娟　姜小苓　王宝石
参编人员：（按姓氏笔画排序）
　　　　万文龙　王廷雨　王园园
　　　　刘用生　张金宝　赵　元
　　　　晁毛妮　程建锋

序

  阅读英文专业文献是培养学生综合英语语言能力和专业素质的重要一环。本书详细介绍了英文专业文献的基础知识、检索、评价、阅读方法和翻译方法，分步剖析了阅读专业研究型论文和综述型论文的各个组成部分，并结合农业科学领域的研究案例（考虑到案例知识产权的保护，书中的案例都是来自编者及其合作者的相关文献），旨在提高读者的英文专业文献的阅读能力和理解能力，以及掌握农业科学领域的专业术语和语言表达方式，从而更好地应对未来的学习和工作挑战。

  本书的出版得到了河南省研究生教育改革与质量提升工程项目（YJS2023JC16、YJS2022ZX22 和 YJS2024S220）和河南省高等教育教学改革研究与实践项目（研究生教育类）（2023SJGLX008Y 和 2023SJGLX056Y）的支持。在本书的编写过程中，还得到了中国农业科学院、河南科技大学、中南林业科技大学、河南农业大学、河北北方学院等单位同行的支持和帮助，在此向他们表示衷心的感谢。同时，我也要感谢出版社的编辑和制作团队，他们认真负责地完成了本书的制作工作，为读者们提供了高品质的阅读体验。

  本书的理论具有广泛性和通用性，适合所有专业的本科生、研究生和研究人员阅读。书中的案例适合作物学、生物学（植物学、遗传学）、园艺学、生态学等学科的学生、教师和研究人员阅读，读者可根据自身专业背景选择适合自己的内容进行阅读。相信通过阅读本书，读者们将会在英语阅读和理解方面取得显著进步，以便更好地掌握农业科学领域的知识和技能，为未来的发展打下坚实的基础。

编 者

2024 年 4 月

目 录

第一章 专业文献基础	1
第一节 专业文献简介	1
第二节 专业英文文献的语言特征	3
第二章 文献类型概论	14
第一节 学术期刊	14
第二节 专著	35
第三节 会议文件	43
第四节 学位论文	48
第五节 科技报告	53
第六节 政府出版物	56
第七节 专利	60
第八节 标准	65
第九节 科学新闻	69
第十节 产品规格书	72
第三章 专业英文文献检索与评价	76
第一节 专业英文文献检索	76
第二节 专业英文文献检索的评价和分析	84
第四章 阅读研究型论文	92
第一节 标题和作者归属	92
第二节 摘要	99

第三节　引言 ··· 107
  第四节　研究方法 ··· 116
  第五节　研究结果 ··· 126
  第六节　讨论 ··· 146
  第七节　结论 ··· 154
  第八节　致谢 ··· 161
  第九节　参考文献 ··· 163

第五章　阅读综述型论文 ·· 171
  第一节　综述型论文介绍 ··· 171
  第二节　标题 ··· 174
  第三节　摘要 ··· 176
  第四节　引言 ··· 184
  第五节　主体 ··· 191
  第六节　结论 ··· 199

第六章　专业英文文献的阅读和翻译方法 ·· 207
  第一节　专业英文文献的阅读方法 ··· 207
  第二节　专业英文文献的翻译方法 ··· 214
  第三节　农业科学领域专业词汇的构词法 ··· 218

参考文献 ·· 237

第一章

专业文献基础

专业文献指的是在特定学科或专业领域中，经过科学研究和严谨论证所发表的、具有权威性和可靠性的学术理论或科研成果的文献。这些文献通常由专业学者或研究人员撰写，经过专业编辑和同行评议发表于学术期刊、会议论文集、学位论文等载体上。

专业文献知识对于深入了解某一学科领域的最新进展、理论探索和实践经验等都有很大的帮助，也是开展研究工作不可或缺的重要参考资料。因此，本章在对科技文献进行概括介绍之后，详细介绍了10种类型的科技文献，即学术期刊、专著、会议文件、学位论文、科技报告、政府出版物、专利、标准、科学新闻和产品规格书。此外，专业文献还包括了各个学科领域内的相关数据库和信息资源。

第一节 专业文献简介

一、定义

专业文献是指学术出版物，这些出版物报道了自然科学和社会科学以及学术领域内的原创性实证研究和理论工作。

无论哪个科学领域，专业文献的主题都包括科学本身，或科学家发现的思想、事实、规律和分类。专业文献的内涵包括以下几个方面：

（一）科学性

专业文献必须基于科学研究方法进行撰写，确保研究结果准确可靠。

（二）严谨性

专业文献需要遵循学术规范和道德准则，在论证过程中严格把关，尽可能避免出现误导读者的情况。

（三）权威性

专业文献的作者通常是该领域的权威人士或者经过长期积累具有深厚研究功底的专家学者，因此其研究成果和观点具有较高的权威性和可信度。

（四）可重复性

专业文献需要提供足够的信息和依据，使得其他的研究者可以对其进行重复和验证，从而保证研究成果的可靠性。

二、类型

（一）专业文献根据载体形式的分类

（1）学术期刊。

（2）专著。

（3）会议文件。

（4）学位论文。

（5）科技报告。

（6）政府出版物。

（7）专利。

（8）标准。

（9）科学新闻。

（10）产品规格书。

（二）专业文献根据加工深度的分类

（1）初级文献（一级文献）。包括首次发表的原始科学资料如介绍具有新颖性科学研究的论文、会议文件、专利、科技报告、标准等。

（2）二级文献。包括综述型论文和书籍，是对一级文献的总结和综合。

（3）三级文献。包括教科书、百科全书和供公众广泛阅读的有关作品。

第二节　专业英文文献的语言特征

专业文献作为一种为了满足书面学术交流和科学研究需求的正式作品，其语言特征主要在于精确性、严谨性和逻辑性，在文体、形态、句法等方面呈现出独特的语言特征。

一、文体特征

在文体上，专业文献是一种正式的书面作品，是一种通往非常重要主题的严格途径，因此要避免模糊和口语化的表达。专业文献的主题是客观的事实或问题，其分析是基于相关的研究数据，而不是主观臆断，专业文献的讨论和结论是基于具体的试验或调查，通常使用第三人称和被动语态来保持客观公正性和科学专业性。

（一）正式性原则

专业文献写作的正式性原则是在保证准确性和完整性的前提下，使用符合学术规范的格式、风格和语言来表达自己的观点，以体现专业性和严谨性。以下是一些实现专业文献正式性原则的方法：

（1）采用正式的学术语言和词汇，避免口语化、俚语化等不适合学术场合的语言。

（2）对重要概念、术语等进行准确定义和解释，确保读者理解无误。

在专业英文文献中，词语的选择通常被认为是局限于正式和标准化的语域，而不是非正式和口语化的语域。以下是一些例子（表1.1）。

表1.1　专业英语文献中非正式和正式词语举例

非正式	正式	非正式	正式	非正式	正式
about	approximately	get	obtain	try	endeavor
ask	inquire	give	accord	use	utilize
balance	homeostasis	have	possess	as a result	consequently

(续表)

非正式	正式	非正式	正式	非正式	正式
begin	commence	help	facilitate	by which	whereby
buy	purchase	produce	yield	go with	accompany
careful	cautious	quick	rapid	in the end	eventually
change	transform	rich	affluent	in which	wherein
end	conclude	say	remark	put together	aggregate
enough	sufficient	so	thus	recurring theme	motif
finish	complete	stop	inhibit	sugar breakdown	glycolysis

（二）简洁性原则

在专业英文文献中，作者普遍倾向于使用有效和精确的专业术语，而不是使用冗长、模糊或不必要的词组，这一原则也被称为词语的简洁性原则（经济性原则）。

专业文献写作的简洁性原则是指作者应该用尽可能少的语言来表达自己的观点，避免不必要的冗长、重复或无关内容。简洁性原则可以帮助读者更快地理解和接受文献内容，提高阅读效率。以下是一些实现专业文献简洁性原则的方法：

（1）确定论文主题和中心思想，避免离题或赘述，切勿浪费篇幅。

（2）使用简单、明确的语言，摒弃复杂、模糊的词汇和句式，使论文更加流畅。

（3）避免使用过多的修饰语和形容词，使描述简洁明了。

（4）尽量使用简短的句子和段落，以便于读者阅读和理解。

（5）采用有力的结论和总结，概括论文主旨和贡献。

（三）直接性原则

专业文献的直接性原则是在保证准确性的前提下，使用简单、清晰、直接的语言来表达自己的观点，避免使用冗长、复杂或晦涩难懂的句式和术语。直接性原则可以帮助读者更好地理解和接受文献内容，提高知识传播和学术交流效果。以下是一些实现专业文献直接性原则的方法：

（1）使用简洁、明确的语言来表述观点，尽量避免使用冗长或复杂的句式和术语。

（2）尽可能使用常见的术语和词汇，避免使用过于生僻或模糊的词汇。

(3) 采用适当的框架结构和标点符号，使论文层次清晰，段落分明。

(4) 在文献中添加图、表、公式等辅助信息，以更加直观、简单的方式展示数据和结果。

以下是直接性原则的一些例子。

conducted an investigation into→studied

not...any→no

not...many→few

not...much→little

"The study found a statistically significant difference between the control group and the treatment group in terms of plant height." 可简化为 "The study showed that the treatment group had significantly lower plant height than the control."

"The research team conducted an investigation into the effects of climate change on plant growth in arid regions." 可简化为 "How climate change affects plant growth was studied in dry areas."

（四）客观性原则

专业文献的客观性原则是指作者应该尽可能地使用客观、中立和准确的语言来表达自己的观点，避免使用主观色彩过强、情感化或武断的表述。客观性原则直接关系到研究结果和结论的可靠性和信任度。以下是一些实现专业文献客观性原则的方法：

(1) 使用客观的事实和数据来支持观点，尽量避免使用基于主观偏见或情感的推断或臆测。

(2) 避免使用赞美词或贬低性语言，使用中立、准确的词汇来描述事物。

(3) 通过对不同观点的讨论和分析，给予读者全面、客观的了解，避免片面或武断的结论。

(4) 在说明方法、试验、数据处理等方面，采用详细、准确的描述，使读者可以重复试验或验证结果。

(5) 引用他人的观点或结果时，需要遵循引用规范，并附带相关出处信息，以保证论据来源的可靠性和透明度。

总之，专业文献的客观性原则是科学研究的基础之一，要求作者尽可能准确、中立、客观地表达自己的观点，并给予读者全面、客观的信息。只有这样，读者才能更好地理解和接受文献内容，提高知识水平和创新能力。

（五）非语言表达的使用

专业文献中的非语言表达（Non-verbal Language），主要包括图表、符号、公式和照片等。非语言表达在专业文献中具有很大的优势，能够帮助作者用规范清晰的方式高效传达信息，提高可视化效果和阅读体验。

二、形态特征

专业文献中的语言形态特征是指在词汇、语法和逻辑结构上具有的一些特殊特征。这些特征是为了保证专业文献的准确性、严谨性和规范性，以满足学术交流和传播科学知识的需要。从形态上看，专业文献的特点是使用科技术语、名词化和限制模糊语。

（一）科技术语

科技术语（Technical and Scientific Jargon，TSJ）或术语（Terminology），是指特定领域中的专业用语，往往具有精确、准确的定义，能够有效地传达相关概念和信息，并在该领域内得到广泛使用，但其他领域的人可能不太容易理解。

根据"技术性"的程度，TSJ可分为通用TSJ和TSJ两大类，前者的数量比后者大得多。

1. 通用科技术语

通用TSJ是指可以在不同的领域使用，但具有不同含义的词，如表1.2。

表1.2　通用科技术语在不同领域的使用举例

通用术语	中文释义	使用领域
culture	文化；文明；修养	社会学
	栽培；养殖	农业科学
	（微生物、细胞、组织等的）培养	生物学
graft	（植物上）嫁接	植物学
	（动物上）移植	动物学、人类医学
	辛苦地工作；贪污	社会学
stock	股份，股票；库存	经济学
	树干；血统（家族）；家畜	生物学
	枪托	军事学

2. 术语

通常仅限于给定领域的单词，如表1.3。

表1.3 特定术语在给定领域的使用举例

术语	中文释义	使用领域
loanword	外来词	语言学
mitochondria	线粒体	生物学
heterosis	杂种优势	生物学

（二）名词化

名词化（Nominalization）是指把动词、形容词或从句转化为名词或名词短语的过程，在一定程度上增强文本的严谨性和准确性。此外，名词化可以使论文更加简洁，这也使论文更加正式。

学术英语名词化一般可分为词汇名词化（Lexical Nominalization）和小句名词化（Clausal Nominalization）。

1. 词汇名词化

词汇名词化有三种类型，即复合型、后缀型和转换型。

复合在这里指的是通过连接两个或两个以上的单词来创造名词的构词过程。以下是一些例子：

cytoplasm（源于cyto和plasm）

greenhouse（源于green和house）

photosynthesis（源于photo和synthesis）

rootstock（源于root和stock）

sweetcorn（源于sweet和corn）

后缀化是指在形容词或动词后面加上后缀构成名词的方式。

由形容词构成的名词的例子：

active（源于activity）

defensive（源于defense）

deficiency（源于deficient）

potentiality（源于potential）

significance（源于 significant）

tolerance（源于 tolerant）

independence（源于 independent）

intensity（源于 intense）

由动词构成的名词的例子：

development（源于 develop）

digestion（源于 digest）

expression（源于 express）

hypothesis（源于 hypothesize）

reproduction（源于 reproduce）

treatment（源于 treat）

translation（源于 translate）

转换是一种不改变词形的名词化。这类词常用的有：focus、review、survey、report、attack、preview、advance、attempt、record、limit、progress、detail 等，这些动词也可以用作名词。

2. 小句名词化

小句的名词化可以通过名词性从句、动名词和不定式来体现。

名词性从句和名词一样，在句子中起主语或宾语的作用。主语或宾语从句也可以放在句末，代词 it 放在主语或宾语的位置。以下是一些例子：

Whether grafting can induce inheritable variation（名词性从句作主语）has been a controversial issue for hundreds of years.

I did not understand *what this means*（名词性从句作宾语）.

Notably, *it*（it 作形式主语）is more difficult to find combinations with strong heterosis *via* the CMS 3-line method than *via* the NMS 2-line method.

In actual production, we can take an annual cotton cultivar as the scion and graft *it*（it 作形式宾语）onto the rootstock of perennial cotton to achieve the results of long lifespan with high and stable yield.

动名词短语可以在句子中用作主语或宾语，起到与名词相同的作用。一些独立的句子，有自己的语法上的或逻辑上的主语和宾语，可以变成动名词短语，成为长句中的主语或宾语。例如：

Producing inexpensive hybrid F_1 cotton seeds with high purity and heterosis（动名词短语作

主语) has significance in commercial breeding.

不定式短语可以在句子中作主语或宾语，起着与名词相同的作用，但具有更多的动词特征。不定式短语可以被替换为 it，以下是一些例子：

To smoke too much is not very good for you.

It is difficult *to detect and describe HGT (horizontal gene transfer) events* in eukaryotes, so this phenomenon is sometimes as controversial as graft hybridization and epigenetics, but recently they are flourishing worldwide in the field of biology.

（三）模糊限制语

模糊限制语（Hedging）是指使用某些词语或短语来缓和或降低某个陈述的强度或确定性，以避免过度承诺或错误陈述。

确定作者在多大程度上相信或准备坚持他们所报道或声称的内容是非常重要的。使用模糊语言最基本的目的是：

（1）尽量减少其他学者反对主张的可能性。

（2）使结果能够更准确地报告，例如，某件事没有100%被证明，但被指出并随后被假定。

（3）在声明可能存在缺陷的情况下，谦逊策略得以执行。

（4）符合学术写作领域公认的惯例。

一般来说，情态动词、某些实义动词、某些情态形容词、某些程度形容词和副词、频率副词、某些名词和介绍性短语都有助于模糊限制语的实现。下面进行分类说明。

1. 情态动词

如 can、could、may、might、ought to、should...

Lipids are the major components of the tissue cell membrane system and *may* also be converted into energy and a variety of physiologically active molecules.

The results of this experiment suggest that the new fertilizer *may* improve crop yields.

2. 实义动词

如 assume、believe、estimate、hint、indicate、interpret、seem、suggest、tend...

This *suggests* that upregulation of GAPDH under low-temperature stress increases ATP production and thus supports plant growth.

These results *indicated* that the samples used for protein identification and quantification were robust.

3. 情态形容词

如 possible、probable、feasible、definite、unlikely、abnormal...

However, these results may provide a *feasible* method for studying the possible genetic and epigenetic movement between scion and rootstock in cotton breeding.

4. 程度形容词和副词

如 approximately、enough、extremely、fairly、generally、most、nearly、perhaps、quite、rather、roughly、slightly、somewhat、too、very...

Of the *approximately* 10,000 accessions preserved in the National Cotton Germplasm Collection (NCGC) of the USA, 581 are wild germplasms.

The use of organic fertilizers can *slightly* increase soil fertility in the long term.

5. 频率副词

如 frequently、hardly、occasionally、often、seldom、sometimes、usually...

Under the action of certain proteases and nucleases, they *often* break into many small pieces and fragmented DNA, eventually becoming tubular dead cells.

Most farmers in this region *usually* irrigate their crops using groundwater.

6. 名词

如 assumption、deduction、implication、indication、possibility、suggestion、tendency...

In addition, the long-distance transfer (such as miRNA) and short-distance transfer (chloroplast DNA, mitochondrial DNA, and nuclear DNA) of genetic material between rootstock and scion provide the *possibility* for genetic regulation of cotton growth and development and breeding of new cotton varieties.

7. 介绍性短语

如 it can be argued that..., it can (thus) be concluded that..., it is our view that..., one can assume that..., to our knowledge...

Therefore, it *can be concluded that* Actin can be used as a reference gene in this experiment.

三、句法特征

句法特征指的是英语句子结构和语法规则上的共性和特点。这些特征包括但不限于

第一章 专业文献基础

时态、语态、从句的使用、动词形式等。

在句法上，专业英文文献具有严格的语法结构，大多数情况下是相当统一的。在专业英文文献中有不同的时态，如一般现在时、一般过去时、完成时、将来时和进行时。此外，被动语态、虚拟语气和祈使句的使用频率也很高。

（一）不同时态的用法

1. 一般现在时

（1）一般现在时用于描述已发表的论文或书籍，其结论被认为是当前有效和相关的，不管这些论文或书籍是最近的还是几百年前的。例如：

Coronatine (COR), a plant phytotoxin similar in structure to jasmonate (Zhang et al., 2021; Zhou et al., 2015), plays an important role in regulating plant growth, inhibiting senescence, promoting cell differentiation, increasing chlorophyll content, and resisting low K (LK) stress (Shen et al., 2018; Xie et al., 2015).

（2）一般现在时用于表示一般的真理或事实、一般规律或研究结果所支持的结论。换句话说，用于描述被认为是永远正确的事情（必然事件）。例如：

Potassium (K) is a major plant nutrient, and its deficiency can limit plant growth and development.

（3）一般现在时用于描述一个装置（因为该装置总是以同样的方式工作）。例如：

The temperature gauge gives an accurate reading in all weather condition.

（4）一般现在时也用于陈述研究目的。例如：

This paper presents a joint analysis of mixed genetic model of major gene and polygene of cryotolerance in cotton.

2. 一般过去时

（1）一般过去时用于描述以过去为开始和结束的事情。这一时态主要用于研究型论文的方法和结果部分。例如：

Three representative cotton seedlings were taken from each treatment.

（2）一般过去时用于描述当前工作所依据的先前研究。例如：

Protein digestion was performed according to the FASP (filter-aided sample preparation) procedure (Wiśniewski et al., 2009).

（3）一般过去时用于描述一个事实、一条规律或一个发现，这些事实、规律或发现被认为不再有效和相关。例如：

Unfortunately, the cultivation history of the arboreal cotton in Yunnan was interrupted for a variety of reasons.

（4）此外，一般过去时和上面提到的一般现在时一样，也用于陈述研究目的。例如：

In this study, we used iTRAQ (isobaric Tags for Relative and Absolute Quantification), a high-throughput protein quantification technology to analyze the differential expression of proteomes in the scion and rootstock of grafted cotton seedlings under low-temperature stress, with an aim to understand the molecular mechanism of induced chilling tolerance.

3. 完成时

（1）现在完成时用于描述从过去开始，一直持续到现在的事情。例如：

Several studies have explored and described the defensive processes of plants against biotic (Ali et al., 2014; Delaunois et al., 2014; Yang et al., 2015) and abiotic (Guerra-guimarães et al., 2014; Song et al., 2011; Zhang et al., 2016) stresses by analyzing the apoplast proteomic profile.

（2）过去完成时用于描述过去某个特定时间之前完成的动作。例如：

A large number of grafting experiments had detected that RNA-mediated gene silencing signals could be transmitted between scions and rootstocks (Cerruti et al., 2019; Li et al., 2019; Rishishwar and Dasgupta, 2019).

4. 将来时

将来时用于写提纲、建议和对未来工作的描述。例如：

In the future, it will be necessary to utilize the excellent traits of perennial species and their underlying genes.

It will be interesting to determine whether reported genes or novel genes can play critical roles in the junction formation of cotton grafting.

5. 进行时

进行时态用于描述持续一段时间的动作或状态。例如：

However, some certain wild cotton and sea island cotton plants cannot produce flowers and fruits in the long-day region; thus, they are challenging to be used for breeding.

（二）被动语态的使用

被动语态是用来表达对已经完成的事情的支持。例如：

These morphological and physiological presentations were well evidenced by differentially expressed proteins (DEPs) in the xylem sap.

被动语态用于讨论作为学科知识体系的一部分而存在的背景，独立于当前的作者。例如：

In 1967, Smith and Van-den Bosh first proposed the term "integrated pest management" (IPM), and IPM was formally accepted by the scientific community in 1972 (Ha, 2014).

在研究型论文的方法部分，尤其是材料准备、试验过程等方面，被动语态优于主动语态。例如：Three representative cotton seedlings were taken from each treatment.

（三）虚拟语气的使用

专业英语文献中虚拟语气常用于提出假设、讨论问题、提出意见或建议等，主要使用虚拟语气的时态形式包括过去时和 would/could/should/might + 动词原形。例如：

The new irrigation system might help to increase the efficiency of water use.

Plants are unable to escape from and often suffer from unfavorable environments, so they would be seriously threatened by these stresses unless they have evolved a mechanism to respond to them.

（四）祈使句的使用

祈使句通常省略主语代词，在句首使用动词原形，主要用于请求、警告、建议或/和下达命令或指示，这一语法特点主要体现在研究型论文的试验方法部分和产品规格书中。例如：

Fill in a tube with tap water, and then heat the tube to 100℃.

 课后练习

与在同一研究领域学习的同学以小组为单位合作，试着找出一些适用于自己研究领域的特定术语。

第二章

文献类型概论

第一节 学术期刊

一、定义

学术期刊是一种旨在推动科学进步的定期出版物，报道新的研究成果。作为新研究的介绍、展示、评价和讨论的平台，通常由学术团体、研究机构或出版社出版。

创办学术期刊的目的是促进学术交流和知识共享，以推动某一特定领域内的研究、理论和应用的发展。其具体目标包括：

（一）发布新颖的研究成果

学术期刊为研究者提供了一个向同行展示其研究成果的平台，从而提高该领域内的研究水平和研究者的科学素养。

（二）推广新观点和技术

学术期刊通过发布关于学术领域中最新趋势和发展的论文，为相关领域的学者和专业人士提供最新的信息和观点，促进学术界的交流和创新。

（三）建立学术交流桥梁

学术期刊将不同地区、不同背景的学者联系起来，促进学术交流和沟通，增强学术合作的可能性。

（四）提供可靠的学术资源

学术期刊所发布的论文通常经过同行评审，确保其质量、可靠性和权威性，因此成为学术研究和教学中有价值的资源。总之，学术期刊的创建和发展旨在为学术研究和教学提供支持，促进知识交流和共享，是学术界不可或缺的一部分。

二、期刊论文的类型

学术文献有不同的类型，其中，一些是原创性研究，一些是基于其他已发表的作品的派生文献。了解可以在期刊上发表的不同类型的论文是很重要的，并不是所有的期刊都发表每一种论文。因此，大多数期刊出版商提供给作者准确和具体的投稿指南，来介绍出版的论文类型。

学术期刊上所载的论文类型多种多样，除了常见的原创研究型论文、综述型论文、研究简报、展望、观点和评论论文，学术期刊上还会出现其他类型的论文，例如科技新闻、案例研究、书评、访谈、译文、读后感等。下面详细介绍最常见的4种期刊论文类型。

（一）原创研究型论文

原创研究型论文（Original Research Article）是指一篇介绍研究者进行的原始试验和研究结果，并对其进行分析和解释的学术论文。其主要特点包括：

（1）独立性：原创研究型论文应当是研究者自主开展研究的成果，不具有抄袭或剽窃他人成果的行为，体现了其独立思考和创新精神。

（2）创新性：原创研究型论文应该有新的、前瞻性的贡献，如发现新现象、提出新理论、解决新问题等。

（3）可重复性：原创研究型论文必须能够呈现具有可重复性的试验设计和数据处理方法，以保证研究结论的准确性和可靠性。

（4）学术性：原创研究型论文应当遵循学术研究规范，采用科学、严谨的态度对研究对象进行描述、分析和解释，同时以客观的方式揭示研究的局限性和不足之处。

（5）同行评审：原创研究型论文通常经过同行评审，以确保其研究设计和数据分析的可信度和可重复性。

（二）综述型论文

综述型论文（Review Article）是指对某一领域内的研究成果、理论和方法进行总结、评价和归纳的学术论文。综述型论文是用来调查和总结以前发表的研究，而不是报告新的事实或分析。综述型论文也被称为概述论文。专门发表综述型论文的学术出版物被称为综述期刊。

综述型论文可能包括背景信息、最近的主要进展和发现、研究中的重大空白、当前的争论和未来的研究课题。其主要特点包括：

（1）全面性：综述型论文应该覆盖所涉及领域内最新、最重要的研究成果，并结合相关文献进行梳理，展现该领域的研究发展趋势和前沿问题。

（2）对比性：综述型论文应该对已有的研究成果进行横向和纵向比较，归纳出优缺点，并就不同研究观点提供自己的解释和看法。

（3）系统性：综述型论文需要按照规律性的逻辑顺序进行组织，以达到系统阐述各种观点的目的，同时使读者可以快速而清晰地了解该领域的研究现状和未来方向。

（4）与时俱进：由于学科的发展日新月异，综述型论文应该对近期发表的研究成果进行追踪，并追踪涉及领域的世界科技研究前沿，保持其信息的准确性和完整性。

（三）研究简报

研究简报（Research Brief）是一种研究性论文，通常较为简短，用于描述作者完成的初步研究成果或者探讨某个新领域的概念与思想。其主要特点包括：

（1）着重突出问题：研究简报论文通常聚焦于解决一个明确的问题，叙述问题的背景、原因、目标等信息。

（2）简练精准：相对于其他研究型论文，研究简报在篇幅和深度上都更加简洁，但需要保证对研究成果和结论分析的描述要清晰、精准。

（3）呈现初步研究成果：研究简报通常描述着重于介绍作者的初步研究成果，而不是长期系统的研究探索。

（4）强调实用性：研究简报通常强调研究成果的应用价值，尤其是针对某个具体领域或行业的实际需求，并且鼓励读者在此基础上进行实践探索。总之，研究简报是一种展示作者初步研究成果及思考方向的研究性论文，简洁明了、实用性强是他的主要特点。

（四）展望、观点和评论论文

展望、观点和评论论文（Perspective，Opinion and Commentary Articles）是学术论文，不需要原创研究。但是，作者需要对这个话题有深入的了解。

展望论文是指对某一重要问题、现象或趋势进行深入分析和探讨的学术论文。展望论文不仅要解释已有研究成果，还需要发掘新知识，提供新观点，并且基于严谨的证据（可能包括原始数据）、系统地分析和推理，得出合理而可信的结论，其主要特点包括：

（1）突出性：展望论文通常选择当前社会热点、重要问题等进行探讨，具有较强的关注度和实际应用价值。

（2）深度性：展望论文的研究深度相对其他类型的论文更高，探讨更多元化和复杂性的问题，并且要求作者有扎实的理论基础和坚实的研究能力。

（3）创新性：展望论文不仅要解释已有研究成果，还需要发掘新知识，提供新观点。

（4）合理性：展望论文得出的结论必须基于严谨的证据、系统的分析和推理，并且与整篇论文的主旨相关联，使读者感到合理和可信。

（5）对话性：展望论文要求与其他观点、研究互动式地对话、交流和分享，以支持和提升其洞见和可理解性。

观点论文是指作者根据自己的认知和分析，表达其对一个假说或科学理论的优点和缺点的看法，并通过理性、合理、有力的论证来支持或反驳该观点的学术论文。观点论文一般都是基于建设性地批评，应该有证据的支持，然而，观点论文不包含未发表或原始数据。观点论文促进科学话语权提升，挑战目前在特定领域的知识状态。其主要特点包括：

（1）主观性：观点论文突出作者的主观思想和判断，强调了作者对所列举的事实和数据的解释。

（2）独立性：观点论文不会受到他人的影响，独立地提出作者本人的观点与判断。

（3）论证性：观点论文需要通过逻辑推理、举例说明、比较分析等方式进行充分、合理的论证，使作者的观点具有说服力。

（4）多样性：观点论文具有多样性，可以涉及社会、科技、文化、政治等各个领域的问题。需要注意的是，观点论文需要通过逻辑推理和相关证据等方式，提高其可信度和可读性，从而使论文具有充分的说服力。

评论论文是对先前发表的论文、书籍或报告的关注或批评，通常使用发现作为行动的号召，或者强调与该领域更广泛相关的几点内容。评论论文不包括原始数据，并且在很大程度上依赖于作者从他或她的个人经验中得出的轶事证据来支持论点。其主要特点包括：

（1）对象性：评论论文的研究对象通常是某一具体作品或事件。

（2）观察性：评论论文需要通过细致的观察和仔细的阅读来提炼作者想表达的意思和思想。

（3）综合性：评论论文应该兼顾多方面因素，如作者背景及时代背景、风格、主题等，全面地分析其内涵和形式。

（4）主客观结合：评论论文需要处理好主客观关系，在尊重作者创作意图的同时，兼顾自己的直觉和体验。

展望论文、观点论文和评论论文的共同点在于都具有学术性质，需要语言准确、逻辑严密、具有可信度和说服力。不同点在于，展望论文注重对一个领域或问题进行全面而深刻的分析和探讨；观点论文着重表达作者对于某个特定问题或者现象的观点和看法；评论论文则是对某一特定作品或现象进行分析与评价。

三、期刊评价工具及评价指标

期刊评价工具可以帮助学术界和科技行业的人们了解期刊的学术质量、影响力和声誉。全球主要的期刊评价工具包括以下几种：

（一）JCR（Journal Citation Reports）

由 Clarivate Analytics 公司推出，是全球使用最广泛的期刊评价工具之一，其核心指标为 Impact Factor。

（二）Scopus

由 Elsevier 公司研发，也是一个综合性的学术数据库和期刊评价工具，其核心指标包括 CiteScore、H-index 等，并推出了 SCImago 期刊排名。

（三）Dimensions

由 Digital Science 公司推出，整合了大量学术资源和数据，提供非学术引用指标 Altmetric，为科研影响力的评价工作提供了非常独特和有价值的维度。

总之，这些期刊评价工具都具有自己的特点和优势，使用者可根据具体的需求和目的选择适合的工具和评价指标来进行期刊评价和排名。

期刊指标（Journal Metrics）是学术界广泛用于评估学术期刊影响力和质量的指标，旨在反映期刊在其领域内的地位、在该期刊上发表论文的相对难度以及与之相关的声望。下面介绍 10 种常用的期刊评价指标。

（一）期刊影响因子

期刊影响因子（Journal Impact Factor，JIF），简称影响因子（Impact Factor，IF），是一种用于衡量期刊学术影响力的指标，随 JCR 一起发布。IF 被广泛用来评估期刊在特定年份内所发表的论文在后续两年内被引用的数量，以此反映该期刊对学术界的贡献和影响。IF 的计算方式为：将某期刊在连续两年内发表的被引用的总次数除以该期刊在同一期间内发表的 article 和 review 数量，得到每篇论文的平均被引用次数；然后再将该期刊的平均被引用次数除以该领域所有期刊的平均被引用次数，得到该期刊的影响因子。IF 经常被用作本领域内期刊相对重要性的代表，IF 较高的期刊往往被认为比较重要。

IF 作为期刊评价的标准之一，具有以下优点：

（1）IF 是一个普遍公认的指标，得到了广泛应用。

（2）IF 能够快速地反映期刊在学术界中的声誉和影响力，并成为许多作者选择投稿期刊的参考。

（3）IF 相对容易计算，方便对期刊进行排名。

不过，IF 也存在着以下缺点：

（1）IF 只反映了某期刊在特定时间段内（包括）的影响力，不能全面反映其长期的学术影响力。

（2）容易陷入"以刊评文"（以刊物的 IF 来评判该刊上所有论文的价值和水平）的误区，因为 IF 只考虑了被引频次而未考虑论文质量和学术影响的范围。

（3）IF 存在领域差异和学科扭曲等问题，可能会导致某些学科的期刊受到不公正的评价。

（4）部分期刊为追求高 IF 而采取投稿限制（例如论文篇数限制、受邀投稿、特定领域和地点限制）、多发综述（因为综述的被引频次相对会比研究性论文的被引频次多一些）、倾向于多发某些研究主题而忽视其他领域的优秀论文、发表评论和简短通讯以吸引更多的读者等措施，可能对学术研究造成一定的负面影响，最终也可能会削弱期刊作为学术媒介的功效。因此，在使用 IF 进行期刊评价时，需要综合考虑其优点和缺点，并结合其他评价指标进行综合评估。

由 Eugene Garfield 和 Irving Sher 于 1963 年创造的"期刊影响因子"概念，自 1975

年首次发布期刊影响因子和他引率（Cited-only Journal Impact Factor）以来，针对其缺点，JCR 不断对其优化和补充。比如，1991 年，推出了即时性指数（Immediacy Index）；1997 年，开始对他引率进行更严格的计算；2004 年，推出了去除自引的影响因子（Journal Impact Factor Without Self Cites）；2009 年，接连推出了 5 年影响因子（5-Year Journal Impact Factor）、特征影响因子（Eigenfactor）和论文影响力（Article Influence Score，AIS）；2014 年，首次提供期刊的自引率（Self-citation Rate）指标；2015 年，又推出了期刊影响因子百分位（JIF Percentile，JIFP）。

（二）CiteScore

CiteScore 指标由 Elsevier 公司于 2016 年 12 月推出，以响应学术界对影响因子的诟病，为学术期刊的引用提供更广泛、更透明的视图。

与期刊影响因子类似，CiteScore 也是以期刊发表的论文被引用次数为基础进行计算。不同的是，CiteScore 同时计算了在当前年份之前 3 年内的引用次数，并将其总和除以该期刊所发表的论文总数量，得到每篇论文的平均被引用次数。由于 CiteScore 并未考虑同领域内的其他期刊，因此可以直接比较不同领域的期刊影响力。

CiteScore 相对于期刊影响因子，具有以下优点：

（1）CiteScore 覆盖范围更广，不仅包括 Web of Science 收录的期刊，还包括 Scopus 数据库中所有已出版的期刊。

（2）CiteScore 考虑了当前年份之前 3 年内论文的总引用次数，反映了期刊长远的学术影响力。

（3）CiteScore 采用了对数函数平均值，避免了一些极端值对指标评价的干扰。

（4）CiteScore 可以直接比较不同领域的期刊影响力，更符合领域间的比较需求。

但是，CiteScore 也存在一些缺点：

（1）CiteScore 仍然只考虑了被引频次而未考虑论文质量和影响范围等其他因素。

（2）CiteScore 仅仅是计算当前年份之前 3 年内的论文被引用次数，并不能很好地反映期刊在更长时间尺度上的学术影响力。

（3）对于一些小众领域或新兴领域的期刊，由于被引用次数相对较少，其 CiteScore 值可能不够准确。

（4）对于一些国家或地区的期刊，由于语言限制或者其他原因，仅有较少的论文被收录在数据库中，这可能会对其 CiteScore 产生影响。

总体来说，CiteScore 作为一个综合评价期刊影响力的指标，具有可解释性强、数据来源广泛、能够直接比较不同领域期刊等优点，但在使用时需要注意其局限性，同时结合其他评价指标进行综合评估。

（三）SCImago 期刊排名

SCImago 期刊排名（SCImago Journal Rank，SJR）是一种期刊评估指标，由西班牙的一个研究小组（SCImago Research Group）开发和维护。SJR 采用一种称为 PageRank 的算法，计算期刊对整个领域的贡献，而不仅是计算期刊的被引次数或引用次数。相比于传统的期刊影响因子，SJR 具有以下优点：

（1）SJR 基于 Scopus 数据库中的论文引用数量，可以反映期刊对学术研究领域的实际贡献水平。

（2）SJR 不仅考虑了被引用次数的数量，还通过参考文献论述的情况来测算论文的质量和影响力。

（3）SJR 使用网络分析的方法计算指标，将研究者与其发表的论文作为节点进行建模，能够更好地体现某些论文、作者、期刊之间的关联性。

（4）SJR 可以根据不同领域的报告传播规律而动态调整指标权重，增强了其适应性。

但是，SJR 也存在以下缺点：

（1）SJR 只依赖于 Scopus 数据库中的数据，可能无法覆盖所有领域或地区的期刊，以及一些非英语期刊。

（2）SJR 并未考虑到某些较老论文所带来的影响，可能忽略了部分期刊的长期影响力。

（3）SJR 所采用的算法较为复杂，难以直观理解和操作。

（4）SJR 使用的是两年引用期范围，不能全面反映期刊和学者的长期贡献。

综上所述，SJR 是一种相对较新的期刊评估指标，具有对期刊影响因素更全面、深入的分析等优点。但同时其也存在一些缺点，需要结合其他评价指标进行综合评估。

（四）单篇论文的来源标准化影响

单篇论文的来源标准化影响（Source-Normalized Impact per Paper，SNIP）是一种用于评估期刊平均引用影响的指标，方法是将期刊被引频次除以期刊所在学科领域中所有文献的平均被引频次，因此修正了学科领域之间引用实践的差异，从而允许对引用影响进行更准确的领域间比较。该指标计算公式为 $SNIP = RIP/(R/M)$，其中 $RIP =$ 每篇论文的原始影响，$R =$ 引用潜力，$M =$ 数据库引用潜力的中位数。由于考虑到了不同领域中的被引频次差异，因此 SNIP 可以更准确地反映期刊的学术价值。

SNIP 的优点包括：

（1）考虑到了不同领域间被引频次的差异，减少了学科之间的影响因素偏差。

（2）可以评估期刊在其领域内的影响力，较为客观地反映期刊的学术价值。

（3）相比其他常用的影响因子指标，SNIP 采用标准化方法，对大量跨学科、多领域的期刊具有更好的普遍适用性和比较性。

SNIP 的缺点主要包括：

（1）SNIP 只考虑了期刊在特定学科领域的被引频次，而忽略了期刊在其他交叉学科领域的影响。因此，某些跨学科或新兴领域的期刊可能无法得到很好的评估。

（2）SNIP 并不能完全代表期刊的学术水平，因为其只是从被引频次这一个角度出发进行评估，而实际上期刊的内容、编委会、审稿流程等也会对期刊的质量产生重要影响。

（3）SNIP 受期刊出版周期长短的影响，出版周期较短的期刊可能受到一定程度的不公平对待。

（五）总引文数

总引文数（Total Citation Count）是指期刊上发表的所有论文被引用的总次数，是评估期刊影响力和学术地位的一个重要指标。

总引文数的优点包括：

（1）总引文数是一个简单明了、易于理解和使用的指标，可以直观地反映期刊在学术界的受欢迎程度和影响力。

（2）总引文数能够充分考虑论文在长时间内的被引用情况，而不仅是特定时间段内的引用情况。

（3）相比于某些计算复杂的指标，如 Eigenfactor（特征因子，该指标衡量期刊在整个引用网络中的影响力）等，总引文数计算简单方便，可以快速获得。

总引文数的缺点包括：

（1）总引文数只考虑了被引次数，没有对引用的类型、来源、内容等因素进行区分，这可能导致一些低质量、无意义或重复的引用被计入其中。

（2）学科特征和学术领域的差异会对总引文数产生影响，可能导致同样的引文数量在不同学科领域中产生不同的影响力和排名结果。

（3）总引文数无法反映论文的传播和影响方式，即不能体现论文被引用的具体情况、位置和意义，因此在评估期刊影响力时需要结合其他相关指标进行综合分析。

(六) 被引半衰期

被引半衰期（Cited Half-life）是用于评价期刊的一种指标，表示某个期刊从当前年度向前推算，引用数占截至当前年度被引用期刊的总引数50%的年数。

被引半衰期的优点包括：

（1）被引半衰期能够从长远角度反映期刊在学术界中的影响程度和稳定性。一个较长的被引半衰期意味着该期刊的论文质量高、持久影响力强。

（2）相比于单纯的引用次数或者影响因子等指标，被引半衰期更加重视论文的持久影响力。因此可以为学者寻找具有长期学术价值的文献提供指导。

（3）被引半衰期计算方法相对简单和直观，易于理解和使用，并且可以跨学科使用。

被引半衰期的缺点包括：

（1）被引半衰期受样本大小影响较大，尤其是针对小型期刊或新成立的期刊，由于其论文数量少，可能会导致数据不准确。

（2）不同领域间的期刊一般具有各自独特的文化特点和发展模式，因此被引半衰期对不同领域的期刊来说参考意义有限。

（3）被引半衰期只反映了过去的影响力，无法直接预测该期刊未来的发展趋势和地位。

因此，在综合分析多个指标的基础上，结合期刊目前的发展情况和战略进行评价更为准确。

(七) 期刊超越指数

由中国科学院文献情报中心发布的《中国科学院文献情报中心期刊分区表》（以下简称"中国科学院期刊分区"），从2022年开始，其分区指标不再采用之前基础版的"3年平均影响因子"，而是替换为"期刊超越指数"（升级版）。

期刊超越指数（Field Normalized Citation Success Index，FNCSI）的计算原理是：随机从一本期刊中选择一篇论文，该论文的引用次数大于从其他期刊随机抽取的一篇相同主题、相同文献类型的论文引用数的概率（50%意味着两本期刊做得一样好）。

期刊超越指数避免了影响因子所导致的以下几种情况：

（1）一篇超高被引论文拉高JIF整体均数。

（2）冷门学科备受冷落。

（3）均值指标容易被人为操控。

（4）期刊学科交叉性无法体现。

（八）综合学术诚信风险指数

综合学术诚信风险指数（Comprehensive Academic integrity Risk index，CAR 指数）是一种用于评价期刊学术诚信风险的指数，该指数综合了期刊本身特征、同类期刊的表现以及已知学术不端行为等多种因素，为读者、作者和科研机构提供一个可靠的参考。CAR 指数是一种计算复杂、权威性高且可靠性强的期刊评价工具。

CAR 指数 = 撤回文章数（R）+ Pubpeer 等平台曝光文章数（P）+ FigCheck 检测出的问题文章数（F）的总和，除以该期刊当年发表文章数（N）的比值，即 CAR =（R+P+F）/N。CAR 指数可以作为部分国际出版商对期刊学术诚信负面相关事件的一种惩罚措施，根据 CAR 指数高低分级，把低于 5% 设定为低风险，5%～10% 为中风险，大于 10% 为高风险。如果 CAR 指数超过 10%，代表该期刊存在较大的学术诚信风险，可能面临被剔除 SCI 收录等风险。因此，实际上 CAR 指数有助于鼓励期刊编辑和出版商们保持出版期刊的学术诚信，同时也为广大读者提供了一个重要的参考指标。

CAR 指数与传统的 IF 不同，IF 只是反映了期刊文章被引频次的情况，并没有综合考虑其他因素，如期刊编委会和作者的影响力等。因此，CAR 指数与 IF 相比，更能全面反映一个期刊的学术诚信风险。

需要注意的是，CAR 指数并不能完全反映一个期刊的学术诚信情况，仅仅是提供了一个基于数据的参考。例如，尽管部分期刊的风险显示数据为 0，还是被剔除 SCI，这种情况其实与该指数的计算方法有关，因为目前 CAR 指数仅包括撤回+曝光+图片重复文章，所以有部分因其他原因（比如自引率高、发文异常、论文质量下降等）被剔除的期刊是无法被查询到的。此外，CAR 指数还在不断发展和完善中，未来随着研究的深入，CAR 指数也可能会发生变化。

（九）期刊分区

目前，期刊分区的方法主要包括以下两种：

1. 基于引文数和影响因子的指标

基于引文数和影响因子的指标，比如 JCR（Journal Citation Reports）期刊分区就是根据期刊论文在特定年份内被引用次数和影响因子等指标来评估和分类期刊。JCR 将收录的期刊分为 176 个不同学科类别，每个学科类别按照期刊影响因子的高低，划分了 4 个区：影响因子排在前 25% 的期刊为 Q1 区，排在 25%～50% 期刊为 Q2 区，排在 50%～

75%的期刊为Q3区，排在75%之后的为Q4区。这种方法有助于评估期刊的影响力和知名度，但也存在可能会忽略其他重要的因素的影响，如期刊的审稿质量、编辑政策、出版周期等。

2. 基于复合指标体系

基于复合指标体系，比如中国科学院期刊分区综合考虑了期刊的学术质量、影响力、管理质量、审稿效率等多个方面的指标进行综合评价和分类。中国科学院文献情报中心将科学（SCIE）、社会科学（SSCI），以及新兴来源（ESCI）引文索引数据库收录的中国期刊（自然科学+社会科学）归属到18个大类学科，期刊超越指数排在各大类学科前5%的期刊为一区、6%~20%为二区、21%~50%为三区，其余的为四区。这种方法相对于仅基于引文数和影响因子等指标的方法更加全面、客观，但也需要考虑指标体系的合理性和是否涵盖了可能影响期刊质量的所有因素。

此外，中国科学院期刊分区还发布了TOP期刊。只要期刊满足以下任意一点，都可以说是TOP期刊。

（1）在大类学科中一区的期刊基本会被直接认定为TOP期刊。

（2）一般大类学科二区期刊前两年的总被引率位于前10%，同样会被认定为TOP级期刊。这种情况比发文量比较大的二区期刊被认定为TOP级期刊的概率要大，毕竟发文数量越多，被引用的次数也就越高。

（3）如果期刊在行业中有很高的权威，且经同行评审评议后，同样也会被列入TOP期刊行列之中。

（4）从2016年起，小类学科的一区期刊也会被认定为TOP期刊。

（十）掠夺性期刊

掠夺性期刊（Predatory Journals），也被称为伪科学期刊或骗子期刊，是指那些以追求经济利益为目的，缺乏学术品质和严格审稿程序的学术期刊。这些期刊通常以与正规期刊相似甚至相同的名称发送电子邮件引诱作者投稿，以快速出版和宽松的审稿流程吸引学者。但是，掠夺性期刊往往不会进行严格的同行评议，或者只是进行表面性的审稿，以获取更多的出版费用。掠夺性期刊的特点包括：

1. 缺乏学术声誉

这些期刊往往没有被广泛认可或被列入学术索引数据库中。

2. 缺乏严格的同行评议

掠夺性期刊通常没有进行严格的同行评议，或者只是进行表面性的审稿，甚至有些可能根本没有进行审稿。

3. 高额的出版费用

这些期刊通常会向作者收取高额的出版费用，有时甚至在文章被接受之前就要求作者支付费用。

4. 广告和垃圾邮件

掠夺性期刊经常通过垃圾邮件或其他形式的广告来联系潜在作者，以吸引潜在作者投稿。

掠夺性期刊的存在对学术界和科学研究造成了严重的负面影响。首先，破坏了学术评价体系，使得学术成果的真实质量难以辨别。其次，浪费了研究者的时间和资源，因为研究者可能会将宝贵的研究成果发表在这些没有学术价值的期刊上。最后，对学术声誉造成了损害，因为一旦文章发表在掠夺性期刊上，该文章的学术价值和可信度会受到怀疑。

为了应对掠夺性期刊的问题，学术界和科研机构采取了一系列措施，包括建立黑名单和白名单、提供指导和培训，帮助研究者识别掠夺性期刊，并制定更严格的出版政策和准则。此外，学者们也应该保持警惕，仔细评估期刊的学术声誉和审稿程序，避免将研究成果发表在掠夺性期刊上。

四、识别码

期刊的识别码是由国际标准化组织（ISO）指定的一系列用于标识期刊的数字代码。目前，主要有以下 3 种类型的期刊识别码。

（一）DOI

DOI 是英文 Digital Object Identifier（数字对象标识符）的缩写，是由国际 DOI 基金会（International DOI Foundation）设计和管理的一种数字标识符，可以唯一地标识一个电子文档或其他数字对象。DOI 为注册和使用数字网络上使用的持久标识符提供了技术支持。主要用于识别学术、专业和政府信息，如期刊论文、研究报告和官方出版物。DOI 的目标是"可解析的"，通常 DOI 指的是对信息对象的某种形式的访问。这是通过将 DOI 与指示对象所在位置的元数据绑定来实现的。

与传统的文献引用方式不同，DOI 不仅是针对一篇具体的论文，而且是针对某个数字对象，例如一本电子书、一个数据集、一个图片、一个音频等。DOI 通常由前缀和后缀两部分组成,其中前缀一般为"10.",后面跟着由数字和斜线组成的字符串,如10.1000/123456。

DOI 的主要作用是在网络环境下保持数字对象的持久可访问性和稳定性,以便用户能够方便地找到并引用相关的数字对象。通过将 DOI 嵌入到论文、书籍、网页等数字平台上,用户可以使用 DOI 来跟踪和引用相应的数字资源,而不受资源所在位置、所有者和格式的限制。

总之,DOI 是一项重要的数字标识服务,广泛应用于科研、出版、图书馆、数字图书馆等领域,给数字对象的管理和利用带来了很大的方便和效益。

(二) ISI 文档解决方案 (IDS) 编号

ISI 文档解决方案 (IDS) 编号 (ISI Document Solution Number, IDSN) 是由美国科技信息公司 (Institute for Scientific Information, ISI) 为期刊所设计的一种标识符号。IDSN 是一个包含 5 位数字和字母的序列,可以用来唯一地标识一期刊及其期号。

IDSN 在 Web of Science 等文献检索数据库中广泛使用,用于获取 Document Solution 中文献的全文,同一种期刊同一卷期中所有的论文 IDS 号是一样的。

(三) ISSN

ISSN 是国际标准连续出版物编号 (International Standard Serial Number) 的缩写,是由国际标准化组织 ISO 制定的一种用于唯一标识连续出版物的国际标准,通常用于期刊、报纸、年鉴、杂志等连续性出版物。

ISSN 号码由 8 位数字组成,有一条横线将 ISSN 号码的前四位与后四位分开。ISSN 号码的最后一位为校验位,其计算方法采用了一种特殊的算法,以保证 ISSN 号码的唯一性和正确性,特别有助于区分同名出版物。

ISSN 号码不仅能够帮助读者快速找到并认定某个期刊的身份,而且对于图书馆、文献管理、出版社等机构也具有很大的实用价值。ISSN 号码还被广泛应用于文献索引、文献互借和文献交换等方面,已经成为全球期刊界通行的一项标准。

总之,ISSN 是一项非常重要的期刊编号体系,简化了期刊的鉴别工作,提高了期刊管理的效率和准确度,对于保障期刊信息的传播和利用具有重要的意义。

五、引文索引

引文索引是书目索引的一种,是出版物之间的引文索引,引文索引通过记录被引用文献和引用文献之间的相互关系，来帮助用户查找相关文献、分析学科发展趋势、评估研究成果的影响力等。

（一）Web of Science

Web of Science 是全球著名的引文索引数据库（https：//www.webofscience.com），现由 Clarivate Analytics 公司维护。该数据库通过记录被引用的文献和引证这些文献的论文之间的相互关系，形成了一个庞大的引文网络，提供了各项指标和分析工具，同时也为学术研究、评估、合作等提供有力支持。

Web of Science 主要包含以下 3 个子库：

1. 科学引文扩展库

科学引文扩展库（Science Citation Index Expanded，SCIE）涵盖了自然科学领域的高质量期刊和会议论文，包括化学、物理、数学、天文学、地球科学、生命科学等多个学科领域，收录了 9 000 多种期刊，数据可回溯至 1899 年。

2. 社会科学引文索引库

社会科学引文索引库（Social Sciences Citation Index，SSCI）涵盖了社会科学及相关领域的期刊和会议论文，包括政治学、经济学、心理学、教育学、传播学、管理学等 12 个子领域，收录了 3 000 多种期刊，数据可回溯至 1956 年。

3. 艺术与人文科学引文索引库

艺术与人文科学引文索引库（Arts & Humanities Citation Index，A & HCI）涵盖了人文学科领域的期刊和会议论文，包括艺术、历史、语言学、音乐、哲学、宗教等多个学科领域，收录了 1 700 多种期刊，数据可回溯至 1975 年。

除了以上 3 个子库，Web of Science 还提供了 Conference Proceedings Citation Index（CPCI）、Emerging Sources Citation Index（ESCI）、Conference Proceedings Citation Index-Social Science & Humanities（CPCI-SSH）、Index to Scientific & Technical Proceedings（ISTP，1980—1995）等子库和工具，以满足不同领域研究者和管理者的需要。

总之，Web of Science 引文索引数据库是全球范围内广泛应用的学术资源平台之一，Web of Science 为学者提供了高质量的文献检索、引文计数、影响因子、作者合作等各项指标和分析工具，对于促进学术交流、评估和发展具有重要意义。

（二）Scopus

Scopus 是 Elsevier 公司推出的跨学科引文索引数据库（http：//www.scopus.com），

是全球最大的引文索引数据库之一，收录了全球超过 25 000 种科技、医学、社会科学、艺术等众多领域的期刊和会议论文，包括来自 150 多个国家和地区的文献。以下是 Scopus 引文索引的特点：

（1）覆盖面广：Scopus 涵盖了各种学科领域的文献，包括科技、医学、社会科学、艺术等。与其他类似平台相比，其覆盖面更为全面，可用于不同学科领域的研究。

（2）引用信息完备：Scopus 对每篇文献都进行了深入的引用分析，可以追溯到该文献被其他文献引用的情况，并提供了多种引用指标，如引用次数、H 指数等，方便评估论文质量和学术影响力。

（3）数据准确性高：Scopus 的数据来源主要是从期刊出版商和会议组织处直接获得，因此其数据质量和可靠性较高。此外，其对文献分类和索引的处理也相对更为精确和细致。

（4）检索与分析工具强大：Scopus 提供了丰富的检索和分析工具，可以根据关键词、作者、文献类型等多种条件来定制搜索。同时还可以通过其高级引用分析功能对文献进行引用分析、共被引分析、关键词频次统计等，方便用户获得更深入的信息。

总之，Scopus 是一款功能强大、数据质量高、覆盖面广泛的学术引文索引数据库，对学术研究人员、科学家等进行学术研究、评估学术成果影响力等方面有重要作用。

（三）Dimensions

Dimensions 是一家新兴的引文索引数据库（https://www.dimensions.com），由 Springer Nature 子公司（Digital Science）开发，目前以其领先的技术优势和创新性的数据系统备受关注。Dimensions 收录了超过 1 亿份文献并能够实现全文检索，同时也提供了精确的引用分析和评价工具，适用于科学研究、学术评估、政策制定等方面。其特点包括：

（1）引用分析：Dimensions 通过整合论文、预印本和会议论文等高质量学术资源，通过引文分析技术将这些高质量学术资源链接在一起，为用户提供更全面的研究视角。

（2）大规模数据集：Dimensions 收录了 60 多个领域超过 1 亿篇学术论文，并持续增长。相比传统数据库，Dimensions 具有更大的学术覆盖范围和更丰富的数据集。

（3）开放数据：Dimensions 提供免费 API，可以让用户快速查找论文、作者、期刊等信息，也可以方便地将这些数据用于其他应用程序或自己的研究中。

（4）可视化分析：Dimensions 内置可视化工具，可以通过图表和地图等方式帮助用户更加直观地分析数据和研究趋势。总之，Dimensions 是一个功能齐全的学术搜索引擎，提供了广泛的学术资源、高度可视化的分析和开放的数据接口，为学者提供了许多有价值的研究工具和信息来源。

六、全球学术期刊出版情况

全球学术期刊出版情况正在迅速发展，对于推动学术研究和知识传播起着重要的作用。以下是一些全球学术期刊出版方面的基本情况：

（一）期刊数量

根据最新数据，全球有超过6万种学术期刊在出版，其中包括英文、中文、德文、法文等各种语言的期刊。

（二）学科领域

学术期刊涉及众多学科领域，如自然科学、社会科学、医学、工程技术等。此外，跨学科领域的期刊也越来越受到关注。

（三）出版形式

随着数字技术的发展，许多期刊现在采用在线数字出版形式，以提供更广泛的读者群体，并且能够更快地传播知识。

（四）开放获取

开放获取已成为近年来学术期刊出版的一大趋势，这意味着论文可以在互联网上免费获取而不需要付费。开放获取期刊的数量正在逐年增加。

（五）国际化

学术期刊成为国际交流合作的桥梁之一，越来越多的跨国出版商和编辑参与到学术期刊的出版中去，促进了全球范围内的知识流动。

学术繁荣离不开科学家、学术机构和国家政策的支持，也离不开学术出版商的贡献。学术出版商的作用和重要性在于：

（一）传播学术成果

学术期刊和书籍是学者们交流和传播研究成果的主要载体。学术出版商通过出版高质量的学术著作，为学者们提供一个广泛的平台去传播他们的发现和想法。这有助于促进学术界的知识交流和合作，推动学科领域的发展。

（二）维护学术标准

学术出版商会对提交的稿件进行严格的审查，确保其中包含了有关的信息、正确的数据以及充分的论证，从而提高学术著作的可信度。通过维护学术标准，学术出版商能够对学术界的发展起到重要的推动和引领作用。

（三）提供优质服务

学术出版商不仅提供学术期刊和书籍，还提供其他服务，比如数据库、在线工具等。这些服务可以帮助学者们更方便地获取和管理信息，提高研究效率。此外，学术出版商也会提供编辑、排版、校对等专业服务，确保学术著作的质量和完整性。

（四）资助学术研究

学术出版商通过出版学术著作来获取经济收益，这些收益会被用于支持编辑、排版、校对等业务以及赞助学术活动和研究项目。这为学者们提供了更多的研究资金和机会，从而促进了学术界的发展。

总之，学术出版商对于促进学术交流、推动学科领域的发展，以及维护学术标准等起着重要的作用。他们提供高质量的学术期刊、书籍及其他服务，使学者们能够更好地传播自己的研究成果并获得更多的研究资源。

以下是全球著名的3大学术期刊出版商：

（一）Elsevier

Elsevier（爱思唯尔）是全球最大的学术期刊出版商之一，总部位于荷兰阿姆斯特丹。该公司拥有超过1 800种学术期刊，涵盖自然科学、医学和社会科学等多个领域。除了学术期刊，Elsevier还出版各种参考手册、数据库和在线工具，如Scopus、Mendeley和ScienceDirect等。

（二）Springer Nature

Springer Nature（施普林格·自然集团）由Springer和Nature两家出版商合并而成，是一个跨学科的出版商。该公司拥有数千种学术期刊，其中包括自然科学、医学、社会科学和人文科学等领域。此外，Springer Nature还出版各种书籍、百科全书和在线工具，如SpringerLink和Nature.com等。

（三）Wiley

Wiley（约翰·威利）是一家跨学科的出版商，其总部位于美国新泽西州。该公司拥有超过 1 500 种学术期刊，涵盖自然科学、技术、医学、人文社会科学等领域。此外，Wiley 还出版各种书籍和在线工具，如 Wiley Online Library 等。

七、全球公认的三大名刊 NCS

NCS 是由三种顶级的科技期刊名称所组成的一个缩略词，包括 *Nature*（自然）、*Cell*（细胞）和 *Science*（科学）。NCS 期刊被认为是全球科学界所公认的最具权威性和最有影响力的期刊之一，这也使得在这些期刊上发表论文成为科学家和研究人员们追逐的目标。这 3 个期刊中的许多论文都被广泛引用，并且对人类生命科学、医疗健康领域等作出了巨大的贡献。以下是这 3 个期刊的简介：

（一）*Nature*

英国自然出版集团于 1869 年开始出版，是全球范围内最著名的科技期刊之一，涵盖了多个科学领域。*Nature* 以其高质量的论文和新闻报道而闻名，特别是在物理学，化学，地球科学，生命科学和医学方面的报道。*Nature* 每年发行 50 期，2021 年的影响因子为 69.504（JCR 排名第 21）。

（二）*Cell*

美国细胞生物学家 Benjamin Lewin 于 1974 年创办，是由克鲁格公司出版的细胞生物学领域权威杂志。*Cell* 在细胞生物学，分子生物学，发育生物学，干细胞研究，癌症研究等方面具有重要地位，并且对科学界产生了深远的影响。*Cell* 每年出版 12 期，2021 年的影响因子为 66.85（JCR 排名第 24）。

（三）*Science*

美国科学家詹姆斯·韦特莫尔于 1880 年创办，是全球最有影响力的科技期刊之一。*Science* 以其高质量的论文，独家新闻报道和社论而闻名，涵盖了多个自然科学领域。*Science* 每周出版一期，2021 年的影响因子为 63.714（JCR 排名第 27）。

八、自然指数统计期刊

自然指数（Nature Index）是由世界著名的科技出版机构自然出版集团（Nature Publishing Group，NPG）从 2014 年 11 月首次发布的一个基于科学论文发表数量（Count）和贡献份额（Share）的指标体系，旨在反映各国、各高校和各科研机构在自然科学领域中的创新能力和科学研究水平，其统计过程具有严格性和客观性，常被广泛用于学术评价和对比研究中。目前，发布这一指数的数据库实时在线版免费向公众开放。

自然指数的核心特点包括以下几方面：

（一）收录期刊数量有限

自然指数仅收录了 82 种（2014 年 11 月开始选定 68 种，2018 年 6 月改为 82 种，2023 年 6 月新增 Health Sciences 类期刊，收录期刊总数增至 146 种）全球顶级的自然科学期刊，具有优秀的影响力和国际声誉，在各自领域拥有广泛的读者、引用量和识别度。这些期刊均由在职科学家所组成的独立小组选出，仅占 Web of Science（Clarivate Analytics）上自然科学期刊的 4%~5%，但在自然科学期刊的总引用量中占近 30%。

（二）数据更新实时

根据知识共享协议，自然指数所涵盖的期刊的每年发文量和被引频次等数据信息可以每年定期在其网站（https://www.nature.com/nature-index）上发布更新，给用户提供及时的、最新的可视化查询和分析工具支持。

（三）涉及领域广泛

自然指数收录的这些期刊不仅来自生物、医学、物理、化学、环境、材料等领域，而且还包括了一些综合性期刊，并且都拥有较为强大的学术影响力和知名度。

入选期刊是 Nature Index 统计的重点和核心之一，通常具有以下特点：

（一）学术质量较高

Nature Index 入选期刊发表的论文均经过严格的筛选和同行评审程序，论文研究内容有深度、有见地、质量上乘，代表该学科领域最新的研究成果。

（二）影响力高

由于这些期刊在各自领域拥有广泛的读者群体和影响力，其论文被引用次数较多，在该学科领域内具备较高的影响力和知名度。

（三）出版质量高

这些期刊对论文发表有较高的要求和标准，包括对学术道德要求、抄袭和剽窃的打击力度、论文创新性和同行评审机制等方面。

（四）国际化程度高

这些期刊主要是国际性期刊，在全球范围内具有一定的影响力和地位，能够反映出不同国家、不同学科领域内的优势和竞争力。

Nature Index 入选期刊的选取标准是通过各种指标数据来评价期刊，并将其分类和排名。入选期刊的指标包括研究型论文的数量、论文被引用的频次、论文作者的国际化程度、发表论文的学科领域等。按照学科分类，自然指数统计的期刊主要包括化学（20 种期刊）、地球与环境科学（16 种期刊）、生命科学（43 种期刊）、物理学（24 种期刊）、卫生科学（64 种期刊）五大类别。因 Nature、Science 等综合类期刊涉及多个学科而被重复统计，因此学科分类中期刊总数超过实际期刊数。

此外，自然指数还会结合其他机构或专家组织推出期刊的综合评价方案，以便更全面、直观地反映各期刊在不同学科领域内的质量和知名度。

需要注意的是，Nature Index 入选的期刊并不意味着其能代表某个领域内最具权威性和声望的所有期刊，尽管会对其学术地位和荣誉产生较大的帮助，但自然指数可能带来竞争性的问题，激发学术机构之间想在排名上占据优势的冲动，而产生不必要的压力和损失。

总体而言，自然指数是一个有争议的、局限性较大的学术评价指标。使用自然指数时需结合具体评价目的分析其数据来源是否符合实际需要。事实上，其他指标，如 IF、SJR、Eigenfactor 等也有类似问题和限制，因此综合考量多种指标可以更加全面客观地评价和比较期刊质量和影响力。

九、农学领域 SCI 期刊出版情况

根据《2022 年中国科学院文献情报中心期刊分区表升级版》，将科学（SCIE）、社

会科学（SSCI），以及新兴来源（ESCI）引文索引数据库收录的中国期刊（自然科学+社会科学）归属到 18 个大类学科 228 个小类学科，农学领域的 SCI 论文主要涉及农林科学、生物学、环境科学与生态学、综合性期刊 4 个大类学科。

表 2.1 是小类学科——农艺学 11 本一区期刊及其 2021 年影响因子（2022 年升级版）。

表 2.1 小类学科——农艺学 11 本一区期刊及其 2021 年影响因子

序号	刊名	ISSN	2021—2022 IF
1	Advances in Agronomy (review)	0065-2113	9.265
2	Agronomy for Sustainable Development	1774-0746	7.832
3	Plant Phenomics	2643-6515	6.961
4	Postharvest Biology and Technology	0925-5214	6.751
5	Agricultural Water Management	0378-3774	6.611
6	Industrial Crops and Products	0926-6690	6.449
7	Agricultural and Forest Meteorology	0168-1923	6.424
8	Field Crops Research	0378-4290	6.145
9	European Journal of Agronomy	1161-0301	5.722
10	Theoretical and Applied Genetics	0040-5752	5.574
11	Crop Journal	2095-5421	4.647

注：数据来源于《2022 年中国科学院文献情报中心期刊分区表升级版》

 课后练习

与在同一研究领域学习的同学以小组为单位合作。试着找到自己研究领域的前 10 名期刊，并提供它们的期刊衡量标准。

第二节 专著

一、定义

专著（Monograph）是指由一位或多位作者，以某个特定主题或问题为中心，详尽地撰写的学术性著作。专著通常会借助严密的逻辑结构、多样的研究方法和充实的案例实证等手段，深入探讨某个领域内的理论、实践、历史和趋势等方面，并承载着作者对相关议题的理性精神及创新思考。

二、类型

从内容与形式上看，专著可以分为以下几类：

（一）理论专著

聚焦于某一学科领域、概念体系、范式模式的基础，依照一定的学科规律，阐明学科理论框架的基本概念、基本原理及其发展历程。

（二）综述专著

从广泛而全面的角度进行剖析和细致解读，将某一领域或特定问题的最新进展、关键问题和最新研究成果梳理拼接起来，并提供对这一领域发展大趋势的总体评估。

（三）研究型专著

在某一领域内开展探讨并提供独创性的见解。这种专著通常包含丰富的案例实证、调查数据、定量或定性分析等内容，通过具体现象和场景来启示和说明理论观点。

（四）年鉴型专著

对某一主题领域内的重大事件、热点话题、行业动态等进行集中归纳和整理，并对其中的争议和趋势进行前瞻性预测和总体评估。这种专著将不同形式的资料汇聚到一起，提供一份精细的年度记录。

根据研究深度、广度和写作风格，一般可以将专著分为一般性专著和学术性专著，其主要区别如下：

（一）研究深度、广度不同

一般性专著的研究领域相对宽泛，重点关注某一个领域普遍认可的基础知识、常见问题或者日常生活中与大众相关联的话题，例如作物种植、病虫害防治、饮食习惯等。而学术性专著则通常针对某一领域或者特定问题的深度探讨，包含具体的理论构建、分析架构、创新思考。

（二）写作风格不同

一般性专著通常采用较为通俗易懂、文字朴实直白的方式，以帮助读者轻松地了解某个领域的基本概念和知识点；而学术性专著则注重阐述难度较高的领域知识，需要使用更加精细和复杂的研究方法，较多运用严密的逻辑和理论分析，需要读者具有基本的相关学科知识。

（三）参考文献引用不同

学术性专著作者通常会注重引用其他研究成果、标注参考文献以及对之前研究进行全面的综述和评估。需要注意的是，一般性专著和学术性专著并不是互相排斥的两类划分方式，在某些领域中两者也可以完美结合。本书主要关注学术专著。

此外，专著还可以按照研究对象进行分类。需要说明的是，专著的分类是为了方便阅读和管理而提出的，且具体分类标准可能因学科领域而异。下面介绍一下国内外主要的图书管理分类方法——图书唯一性标识编码系统和图书分类编码系统。

三、图书唯一性标识编码系统

标识图书唯一性的编码系统是对各种出版物进行标识和管理的技术手段，其重要性和必要性主要体现在以下方面：

（一）唯一性

每个编码都是唯一的，避免同一图书被重复采购或借阅。

（二）信息准确性

标识编码系统可以确保图书馆所拥有的藏书信息更新快速、准确、及时与完整。

（三）效率

操作简单，数据管理更加顺畅，降低了人为错误的发生，提高办公效率。

（四）资源共享

这些编码系统的普及不仅可以促进图书馆之间的合作和资源共享，而且还可以为图书数字化工作提供准确可靠的数字标识符。

目前，主要的图书标识编码系统包括：

（一）ISBN

ISBN（International Standard Book Number，国际标准书号）是对每一本出版物进行全球独立和唯一标识编码的一种系统，是由国际图书出版界制定并推广的一种标准的出版物标识符，由13位数字组成，其中前三位表示语言或地区、中间一段表示出版者代号、最后一位为校验码。

（二）ISSN

ISSN（International Standard Serial Number，国际标准连续出版物编号）用于标识所有期刊、报纸、杂志等连续出版物，由8位数字组成，少量专著也会在连续出版物上出版。

（三）DOI

DOI（Digital Object Identifier，数字对象标识符）是一个具有唯一性的数字串，用于标识网络上的各类电子文献、电子图书和数据集等资源。

四、图书分类编码系统

图书分类编码系统是将图书按照一定的规则和方式进行分组、编目，以便更加方便用户查阅和利用。其重要性和必要性主要表现在以下几个方面：

（一）便于查找和使用

分类后的图书，使得相关的图书被归类到同一个类别中，方便读者快速找到所需要的资料和信息。

（二）提高图书管理效率

分类和编目可以有效地提高图书馆的管理效率，方便图书的存放、借阅、清点、统计等工作，减少重复采购和浪费资源的情况。

（三）面向未来的发展

随着信息爆炸时代的到来，图书的种类和数量也在不断增长。合理分类和编目会帮助图书馆更好地应对信息数量的增加、满足读者不断变化的需求、引导读者发掘更多的知识和信息，同时为图书馆的可持续发展奠定基础。

图书分类编码系统是对图书馆资源进行分类、编号和管理的重要工具。国内外流行的主要图书编码系统包括以下几种：

（一）杜威十进制图书分类法

杜威十进制分类法（Dewey Decimal Classification，DDC）由美国图书馆学家杜威于 1876 年创制，分为 10 个主要类，最后的号码从 000 到 999 作为各类题目的具体代号，可以分类大量相关主题的资料。

（二）通用十进制图书分类法

通用十进制分类法（Universal Decimal Classification，UDC）由比利时图书馆学家 Paul Otlet 和 Henri la Fontaine 于 19 世纪末在杜威十进制图书分类法的基础上继续研发的分类方法。UDC 目前已被世界上几十个国家的 10 多万个图书馆和情报机构采用，已成为名副其实的国际通用文献分类法。该分类法将知识体系分为 10 类，采用小数点后增量式编码，分类精确，可扩展性强。

（三）美国国会图书馆分类法

国会图书馆分类法（Library of Congress Classification，LCC）是美国国会图书馆（Library of Congress，LC）制订并使用的一种分类法，该方法基于学科领域，将所有的学科分为 21 个大类，从 A 到 Z，不同的字母代表不同的学科分类，采用"一个字母（学科）-三位数字（主题）-三个字母（作者）"相结合的方式进行编码，适用范围广泛。

（四）中国图书馆分类法

中国图书馆分类法（原称"中国图书馆图书分类法"）是中华人民共和国成立后编制的一部具有代表性的大型综合性分类法，是当今国内图书馆使用最广泛的分类法体系，简称"中图法"。《中国图书馆分类法》初版于 1975 年，2010 年出版了第五版。《中国图书馆分类法》是以学科分类为基础，结合图书资料的内容和特点，将图书分为马列主义、毛泽东思想，哲学，社会科学，自然科学，综合性图书五大部类，22 个基本大类。

（五）主题分类法

主题分类法是一种基于主题概念而建立的分类法，与传统的基于形式的分类方法不同，主题分类法严格依据文献的主题内容进行分类。如今很多图书馆和数字化资源都采用主题分类法将数据集合起来，以便用户搜索和浏览。

五、学术专著的特点

学术专著作为一种主要的专业文献，主要具有以下5个特点：

（一）学术性

学术专著是专门研究某一特定学术领域的著作，需要有足够的基础体系、深入的研究和详细的解释，来报道该领域的最新理论和实证研究成果。因此，学术性也是衡量专著是否是学术专著的重要标准之一。

（二）独创性

学术专著不仅是对前人研究成果的归纳和总结，而且要在前人工作的基础上发挥自己的思想和见解，展示作者的独特或新颖思路，开创相关领域新方向。因此，专著最显著的特点是在特定领域的创造性，即在科学的分析讨论和归纳的基础上，有创新和创造性的想法和发现。

（三）权威性

专著展示了长期有效的测试结果和深思熟虑的想法，得到各种来源大量证据的证实，并能提供相应的实质性参考资料。此外，学术专著需要经过审查、修改和确认，确保其中的数据、试验结果和理论推导等内容真实可靠，符合科学严谨的要求，因此具有很高的科学性和权威性。

（四）系统性

学术专著通常着重从系统性、全面性的角度讨论研究对象，需要包含观察过程中所涉及的各种相关因素和关键点，以保证读者可以从中获取足够的信息广度和深度，其中包含着几代学者的真知灼见和思考。

（五）超越时间和空间

专著通常包含着人类跨越国界、超越时间的公认智慧，为未来的研究者和学生提供基本的理论知识。

六、学术论文与专著的区别

学术论文和专著在学术研究中扮演着不同的角色，主要区别如下：

（一）研究深度和广度

学术论文通常针对某一具体问题或课题展开系统研究，对问题深入探讨、分析和解决，并且着重关注最新的研究进展。而专著则更多的是以一定范围或领域为主题全面论述，涉及该领域内相关的各项问题和研究成果，并对其历史和发展进行比较全面的综述。

（二）要求和适应对象

学术论文主要针对已具有一定学术背景的读者，追求突出、精准的论证以引起学术界关注，并通过期刊等科学刊物公开发表，获得学术界认可。而专著则更注重广大读者群体的接受，强调语言简洁、条理清晰，力求使大众能够容易地了解其中的内容，对这个主题有基本全面的认知。

（三）出版形式和时限

学术论文通常出版于学术期刊上，期刊会对论文进行审阅和修改，并分别刊载在相应的栏目中完成公开发表。学术论文的出版周期通常较短，且更为灵活，能够及时发布新发现的研究成果。而专著通常以图书出版的形式呈现，一般需要经过较长的编辑、策划，出版时间通常较长。

（四）研究方法和说明要求

学术性论文是采用经典实证法来进行研究，并注重对涉及内容的准确性做出详细解释和合理阐述，需要围绕其研究对象严密论证；而专著的写作可以差异比较大，但一般情况下不再单纯地使用经典工具和技巧，通常也需要准确注释并配有统计数据和各种现象的可能原因和排除方式。

总之，学术论文和专著作为科研领域中的不同类型的文献资料，都有其特定的内在价值和适用范围。选取何种类型文献参考考察体系、所研究领域以及主要目的。学术论文和专著作为两种不同的专业文献，在很多方面是不同的。

七、学术专著的评价

学术专著是自成体系的单卷著作（书）。评价一本学术专著的总体水平，通常需要考量以下几个方面：

（一）作者背景

作者是否具备相关学术背景和经验，是否在该领域内有较高的声望和知名度。

（二）出版社

出版社是否是具有良好声誉的或学术性质的出版社，是否有严格的审核制度、编辑流程以及专业的图书发行管控体系。

（三）引用情况

查看该书的引用情况，即该书是否被其他相关专家学者引用。若被大量引用，则表明该书对于该领域的学术研究具有很高的参考价值。

（四）内容质量

内容是否符合学术研究的要求，有无合理的论证结构，是否有翔实的数据支持，是否合理引用了相关前人的研究成果，学术上是否有创新或者突破点。

（五）学术评价

除了查看读者评价，还应该关注该书是否得到了同行专家的肯定和认可。

总之，一个正规的学术专著应当科学、全面、系统地涵盖该领域的重要问题，基于学术进展与实践，立论严谨而深入，数据和事实的支撑充分且可靠，具有广泛而深刻的影响。同时也需要符合出版社、专家评审团队等对于学术研究书籍的审核要求与标准，体现其学术性和专业性。

课后练习

找一本自己学术领域的学术专著,读一读。与同学分享它是如何体现本单元中提到的学术专著的特点,并对其作一个总体评价。

第三节 会议文件

一、定义

会议文件是指从准备会议到会议结束产生的各种记录、资料和文献,通常包括了与会者之间的谈话记录、演讲稿、展示文稿、决议、报告等。会议文件通常会披露相关领域的最新发现、成果、成就和发展趋势。虽然会议文献中的研究并没有得到充分的认可,但会议文献在一定程度上反映了世界或某些国家在相关领域的研究状况,因此是科学新方向的重要信息资源。

二、会议文件的类型

会议文件包括从会议准备到会后记录的所有印刷材料。根据会议阶段,文件可以分为三大类:

(一) 会议前分发的文件

这类文件包括会议通知、征文通知、作者须知、发言人须知、参会人员登记表、邀请函、初步印刷的会议录、会议筹备情况等。

(二) 会议上分发的文件

这类文件包括日程表、会议论文(摘要)集、会议报告、会议记录、会议决议、会议照片、会议录像和音频资料、开幕词和闭幕词等。

（三）会议结束后分发的文件

这类文件包括会后会议记录、会议文件特刊、会议报告、会议事务和会议促销等。总之，会议文件作为珍贵的信息资源记录了有关会议的重要内容和产生过程，具有重要的历史、文化和研究价值，因此也应该得到妥善保管和利用。

三、学术会议的类型

学术会议是指专业性较强、面向学术研究和学术交流的会议，为会议主题内从事相关研究的学者提供交流研究成果、分享学术经验及开展合作等方面的机会。学术会议可以根据不同的方面分类，如下：

（一）按照规模

学术会议按规模可分为小型、中型和大型。小型会议参与者一般在100人以内；中型会议参会人数通常在100~1 000人；大型会议则往往超过1 000人。

（二）按照领域

学术会议可按照讨论的具体领域进行分类，例如遗传学、生物信息学等。

（三）按照研究对象

学术会议可按照研究的具体对象进行分类，例如小麦、玉米、棉花等。

（四）按照周期性

学术会议也可以按照其周期性进行分类。有些会议每年举办一次，而有些可能是两年或3年举办一次。

（五）按照出版物

学术会议的出版物也可以用于区分，一些会议会制作会议论文集或特别出版物，当然也有一些会议没有出版物。

（六）按照地域范围

学术会议分为国内学术会议和国际学术会议，二者在参与者来源、研究内容、学术水平和会议语言上存在较大差异,特别是在研究内容方面,国内学术会议注重解决当地实际问题,发掘和弘扬本国的学术特色;而国际学术会议则能够促进多领域、多国家的知识交流、资源共享及学术合作。

(七) 按照形式

学术会议还可以根据形式进行分类,包括会议的语言、发言的类型、组织的形式等。

英文中,对不同形式的会议区分较多,以下是一些不同类型的学术会议。

(一) 会议 (Conference)

大型、跨学科、综合性的学术会议,为从事相关研究的学者提供广泛的交流和展示自己研究成果的平台。参与人数众多,一般持续几天。

(二) 论坛 (Forum)

较小规模的会议活动,主题更为特定,有时在学术会议之前或之后安排进行。通常在少数专家的主持下进行深层次的讨论,持续时间可长可短。

(三) 专题研讨会 (Symposium)

较大型、跨学科性的学术会议,面向某个主题或概念进行广泛深层次的深入研究,汇聚了多领域学者的智慧与经验,并以不同形式发布其研究成果,普遍持续多天。与Conference相比较,Symposium一般更狭义特指某一范围。在规模上专题研讨会比Conference小,类似forum,但参与人数较多,而且较正式。

(四) 讲座 (Lecture)

一种专家发表演讲的形式,每人在大约数十分钟内讲述某个话题的相关内容。参加人数一般较多,演讲讨论之间通常会有交互。

(五) 专题报告会/学术讲座 (Colloquium)

一种聚焦于某个具体话题、向参与者提供全面探讨机会的讲座。通常在学术界中被用作介绍独立研究的成果以及推广。与Lecture不同的是,Colloquium强调与听众以相对轻松的方式展开互动和交流,通常涉及特定领域的专家或研究者分享的新成果或观点,而不仅仅是将某个课题呈现给公众。

（六）小型研讨会（Seminar）

一种根据指定课题或议题进行分组的小型会议，重点是对研究进度的深入讨论和学术分享，参与人数有限。

（七）小组讨论会（Panel Discussion）

专家和观众共同参与的小组讨论，以特定研究领域为重点，通过依次轮流发言及日常思想交流来深度探讨课题。小组讨论会通常包括一个主持人，负责指导讨论，有时也会引出听众的问题，目的是提供信息。

（八）分组讨论会（Breakout Session）

分组讨论会是指大型学术会议的一部分，通常在全体会议后召开，由多个小组分别进行具有深度和广度的讨论。每个小组都会针对特定领域或子领域的主题展开辩论和探讨，旨在促进不同参会人士之间对该主题的更深层次理解和交流。通过参与 Breakout Session 帮助参会者更好地了解某个特定话题或问题，并结识同行业内部的专家和同仁。

（九）研习会/实验室研讨会（Workshop）

通常由少数专家主导，通过任务型方式进行问题解决，着眼性强，实践操作性强，目标明确。参与人数较少（通常是几名到几十名），时间一般较短，具有知识传递和技能提升的双重目的。Workshop 可以是会议期间举行的小型会议。

四、参加学术会议的好处

参加学术会议的好处有以下几个方面：

（一）拓宽视野

参加学术会议可以拓宽研究人员的视野，了解最新的研究成果和趋势，接触到不同学科领域的先进技术和理念。

（二）学术交流

参加学术会议可以促进来自不同背景和地区的研究者之间的交流，进行合作探讨，寻求优秀论文谈判和合作机会。

（三）发表论文

学术会议提供了互相发表论文、分享研究成果的平台。

（四）增长见识

窥知行业发展动态以及新兴工作领域信息等。

（五）认识同行

学术会议为研究者们发现别的单位里的真实工作习惯提供一个较好的渠道，也知道公司考虑什么才能为他们挖掘更多潜力。

而参加国际学术会议，除了上述好处，还有以下几个方面的益处：

（一）全球化

参加国际学术会议可以促进科研人员与来自世界各地的学者和专家建立联系，并借助多元化的视角，探讨并解决全球性问题。

（二）跨领域合作

国际学术会议覆盖的领域和专业范围更广，可以让科研人员接触和了解来自不同领域的最新学术成果，从而创造跨领域合作的机会。

（三）冲击力

在国际学术会议上获得口头或海报展示的机会，能够提升研究成果的曝光度，增强学术影响力，为个人职业发展打下坚实基础。

（四）创意汇思

与国际学术会议上与会的专家交流，沟通创意时，理念交融会很大程度上激发出更丰富的新想法。

总之，参加国际学术会议不仅能够获得常规学术会议上所具备的优势，还能够促进科研人员构建全球化视野、跨领域合作和提高影响力等。

课后练习

既然参加学术会议对自己有利,那就需要了解自己研究领域的权威会议。与在同一研究领域学习的同学以小组为单位合作,尽量找一些专门针对自己研究领域的权威会议。

第四节 学位论文

一、定义

学位论文(Thesis/Dissertation)是对大学学位要求的一部分,博士水平的论文通常被称为Dissertation,但Thesis和Dissertation通常被统称为学位论文。

学位论文是本科生或研究生在完成学业时必须完成的一项重要任务,具有如下几个方面的必要性和重要性:

(一)学术能力考核

学位论文对于研究生而言是一项对其学术能力的综合考核。通过撰写学位论文,可以检验研究生的思辨能力、创新精神、综合分析能力、论证能力以及对自己所选专业领域专注深度的理解。

(二)知识传承推广

学位论文的撰写需要针对某一具体问题进行系统深入的研究,并进一步将获得的成果传承给其他人。通过公开发表、宣传和分享自己的研究成果,可以帮助他人了解相关知识,并且激发和提升专业圈内对于此领域的研究兴趣。

(三)职业发展机遇

有一些职业可能会更关注个人的论文研究水平。对于许多学术岗位来说,学位论文是最为重要的工作表现之一,可以影响未来的职业发展方向。具有高质量学位论文的学生，打下了拓展学术界、教育界、公共政策领域等从事有关工作的基础。

（四）科学研究贡献

通过学位论文的撰写与发表，研究生可以为学术界和社会作出创新的贡献，并帮助其他人更好地理解研究所涉及领域和问题的本质，在提升自身的同时也为大家提供了其研究领域的最新概念或洞见等。

二、类型

根据不同的分类标准，学位论文可以分为多种不同类型。从以下几个角度分别介绍：

（一）学科性质

学位论文可以分为自然科学、社会科学和工程技术等不同学科领域。

（二）学位层次

可分为学士学位毕业论文、硕士学位毕业论文、博士学位毕业论文。

（三）专业性质

学位论文可以分为专业性的和综合性的，专业性的学位论文围绕具体学科领域进行深入研究，而综合性的学位论文则要求对多个学科领域进行交叉探究。

（四）研究方式

学位论文可以分为实证研究与理论研究。实证研究通常是基于采集数据，整理与结果分析。理论研究则侧重于对现有学术理论的探索、总结和归纳。

三、格式

学位论文遵循特定的格式要求，英美国家学位论文（尤其是英语国家的博士学位论文）与我国学位论文的格式要求大同小异，通常由以下几个部分组成：

(一) 标题页

包含论文题目、作者姓名、导师姓名、学位类型等信息。

(二) 摘要和关键词

学位论文摘要的字数一般要比期刊论文多一些,博士论文1 000字左右,硕士学位论文摘要的要求是500字左右。概述论文的主要内容和贡献,同时列出3~5个关键词。

(三) 目录

列出论文中各个章节的名称和页码。

(四) 引言

介绍研究的背景和问题、阐述研究目的和意义、提出自己的研究问题并概述论文的结构。

(五) 文献综述或理论基础

如果需要全面回顾前人的研究成果,则这部分是相关工作的详细综述,否则可以涵盖一些基础知识和原理。

(六) 研究方法/试验设计

具体阐述自己采用的研究方法或试验设计,明确数据来源或者参考材料。

(七) 试验结果或数据分析

呈现试验过程中的数据、试验过程曲线图或表格,同时进行定量或定性分析。

(八) 讨论或结论

总结研究结果,进行探究性的讨论或归纳,可能会提出新的理论、方法、技术或方向。

(九) 致谢

感谢所有支持自己学习与研究的人员和单位,并进行相关致谢。

(十) 参考文献

罗列论文中引用的文献，并按照一定格式标准进行排版（如 APA、MLA 等）。

(十一) 附录

包括论文中需要的辅助信息、数据、公式推导过程等内容。

学位论文需要遵循学位授予单位具体的排版格式要求，例如字体、字号、行距、页眉页脚等均有特定要求，常见的格式要求有 APA、MLA 等。此外，有些学位授予单位还可能对页面边距、页码、参考文献的数量、页码展示方式等有更为详细的要求。

四、硕士毕业论文和博士毕业论文的区别

由于学士学位论文的学术参考价值较低，而且一般不会被公共的大型数据库收录，这里主要讨论硕士毕业论文和博士毕业论文。

在英语国家，Thesis 和 Dissertation 的区别会因机构、国家和专业而异，区别包括以下几个方面：

(一) 适用程度

在美国和加拿大，Thesis 通常用于描述硕士毕业论文，而 Dissertation 则用于博士毕业论文。但在英国和其他一些英语国家，这两个术语可以互换使用。

(二) 研究深度和广度

通常认为，博士毕业论文需要更高的研究深度和广度，涉及的主题和问题范围更广泛，对理论与实践的融合探讨也更为深入，要做出学术界领导地位的创新性贡献。硕士毕业论文则相对浅显，通常依据之前的研究成果进行改进或应用。

(三) 论文结构

论文的内容结构可能有所不同。比如，在美国和加拿大，硕士毕业论文通常需要撰写一篇长文论文；而博士毕业论文需要针对领域知识问题进行探究并提出原创性解决方案。

（四）质量标准

由于博士毕业论文必须证明作者在该领域具有权威地位，其质量和知识深度更高。通常要求研究成果在学术界得到广泛认可。

总的来说，在英语国家，Thesis 和 Dissertation 的区别需要依据机构、国家以及具体的专业而定，但通常博士毕业论文需要具备更高的研究深度与广度、创新性，以及论文要求等方面。

五、学位论文和期刊论文的区别

尽管学位论文和期刊论文在某种程度上是相似的，但在某些方面也存在不同。两者之间存在以下区别：

（一）写作目的

学位论文的目的是评估作者学术研究能力，通常是实现学位的授予；而期刊论文则旨在发表对学术和行业发展具有重要性的原创研究成果，以促进学术界和相关行业的发展。

（二）作者数量

学位论文只能由一个作者撰写，而期刊论文可以由多个作者共同撰写。

（三）长度和深度

学位论文相较于期刊论文而言更为冗长和深入，通常包含多个章节，要求作者在研究方法上详细阐述，呈现并证明自己的理论思考和成果，而期刊论文则需要高度概括清晰地呈现研究结论。

（四）摘要方面

学位论文的摘要比期刊论文的摘要要长。学位论文的摘要不仅包括论文的摘要信息，还包括论文的结构，而这点在期刊论文的摘要中往往被省略。

（五）评价方面

除了需要同行评价，学位论文还需要导师评价，学生需要进行口头答辩，以便进一步评估，而期刊论文则需要编辑进行评估。

（六）发布和销售

学位论文一般通过大学图书馆以电子和纸质形式存档，并非商业出版物；而期刊论文则会通过各种学术刊物、商业出版物及互联网等渠道出版和传播。

 课后练习

根据自己的研究领域，找到一份英文硕士毕业论文和一份英文博士毕业论文对两者进行比较，看看在研究深度和研究广度等方面有什么不同。

第五节　科技报告

一、定义

科技报告是描述某项技术研究问题的过程、进展、结果或现状的书面材料。这些报告通常涵盖具体的技术信息和相关文献资料，还可能包括研究的建议和结论，旨在传达完整、准确、系统的研究结果给读者。科技报告是当今科学技术信息的主要来源之一。

科技报告通常是为研究项目的资助方准备的，根据通用标准或资助方设计的标准、逻辑规则和实践经验编写。科技报告被认为是"非档案出版物"，主要是因为科技报告通常不被视为永久存档的学术文献，不论是否修改都可以自由地经过同行评审公开发表。

二、类型

一般来说，科技报告可以从不同的角度分为不同的类型。

科技报告按内容可分为研究成果报告、技术信息报告、技术规范报告、技术评价报告、技术建议报告、技术手册和说明书报告。

根据研究方法可分为试验报告、调查报告和综述报告。

从研究进展来看，科技报告可分为初步报告、进展报告、中期报告和终期报告。

按发布和使用范围，科技报告可分为内部报告和外部报告。

按照传播范围，科技报告可分为绝密报告、秘密报告、保密报告、限制性报告、非保密报告和解密报告。

就正式性而言，科技报告可分为非正式报告、半正式报告和正式报告。

三、特点

相对于其他类型的学术文献，科技报告具有以下特点：

（一）科技报告通常不会经过同行评议的过程

（二）科技报告以最具创新性的研究途径和方法揭示了科技领域的最新发现和成果，对其他研究者具有重要的参考价值

（三）作为一种特殊的专业文献，许多科技报告是在政府的指导下或与高科技相关的。因此，很多都是保密的，只在一定范围内传阅

（四）关于科技报告的出版，每一份科技报告都是独立的，有一个特定的主题，作为单独的特刊出版。但是，同一机构、同一类型等的科技报告需按顺序编号。科技报告编号是科技报告最重要的识别码和查询对象

四、农业领域的科技报告

世界范围内，国际农业机构发布的科技报告包括：

（一）联合国粮食及农业组织（FAO）

发布有关全球农业、粮食安全和食品生产的科技报告，包括有关可持续农业发展和粮食安全的报告。

（二）国际农业研究磋商组织（CGIAR）

是一个国际性的农业研究组织，在全球范围内开展农业科学研究，并发布相关的科技报告，如有关气候变化和种质资源保护等报告。

（三）国际食物政策研究所（IFPRI）

这个以国际视角为主导方向的研究机构致力于帮助提升全球范围内的食物安全问

题。IFPRI 发布大量的科技报告，旨在为促进可持续发展目标作出贡献。

（四）国际热带农业中心（CIAT）

或称波哥大国际农业中心，CIAT 支撑许多经济作物品种的改良，该机构还发布了有关经济作物栽培、种植等方面的科技报告。

（五）国际农业发展基金会（IFAD）

致力于改善发展中国家的农村贫困和营养状况，IFAD 发布报告关注农业发展、食品安全、土地管理等内容。

除了国际农业机构的科技报告，世界上还有大量的科技报告，其中美国农业科学相关的科技报告非常多。就其参考价值而言，以下是几个比较重要的：

（一）USDA 年度报告

由美国农业部（USDA）生成的年度报告，分析了当前的农业形势、问题和主题等方面。

（二）美国农业研究所年度报告

这份报告记录了该机构的活动和研究成果，并展示了最新的突破和未来计划。

（三）食品安全委员会年度报告

这份报告汇总了有关食品安全的最新研究信息并提供了针对不同问题的建议。

（四）农产品出口促进局年度报告

通过分析以前和当前的国际市场，为美国农民指导其如何将最优质的农产品推向海外市场。

（五）气象报告

气象对于农业生产至关重要，研究如何更好地利用气象数据以及采用最新的天气决策支持系统都非常关键。

除了以上报告，还有许多专业性较强的科技报告，例如关于植物、动物遗传学和基因工程、土壤保护与重建、粮食安全等方面的报告，这些报告对于美国农业科技的发展有着很大的推动作用。

课后练习

与在同一研究领域学习的同学以小组为单位合作。在自己的研究工作中,尽量找到自己可以参考的科技报告。

第六节　政府出版物

一、定义

政府出版物是由各级政府、政府机构和部门公开发行的文献资料。政府出版物通常反映了官方对某一特定问题的立场,以及相关政策、法规、措施等信息。政府出版物旨在以权威性和开放性的方式使信息公之于众,因此具有权威性和公共性。

二、类型

政府出版物的形式和类型多种多样,包括但不限于以下几种:

(一) 政府工作报告

政府机构发布的关于特定问题的详细研究报告,如经济报告、环境报告、教育报告等。

(二) 法律法规

如土地管理法、种子法、农药管理条例等各类法律条款和规章制度。

(三) 统计数据

政府部门发布的各类统计数据,如人口统计、经济统计、社会统计等,用于研究和决策参考。

（四）政策文件

政府发布的关于特定政策的文件，如发展规划、行动计划等。

（五）宣传材料

政府部门发布的宣传册、手册、海报等，用于向公众传达政府政策、宣传政府成就和倡导公民责任。

（六）司法文件

包括法院判决、裁定、调解书等，用于记录和解决法律纠纷。

（七）教育教材

政府出版的教育教材，如课本、教辅材料等，用于教育培训。

（八）档案文件

政府机构的档案文件，用于记录和保存政府机构的决策过程和行政管理。

（九）社会服务指南

政府发布的社会服务指南，如医疗保健指南、社会救助指南等，用于指导公众获取相关服务。

（十）审计报告

政府机构或审计机构发布的关于政府财务和行政管理的审计报告，用于监督和评估政府机构的运作情况。

这些形式和类型的政府出版物旨在提供信息、传达政策、记录决策、指导行动，并促进政府与公众之间的沟通和互动。

三、国际农业政府出版物

国际上农业方面的政府出版物很多，以下是其中一些重要的：

（一）Food and Agriculture Organization（FAO）of the United Nations Publications——联合国粮食及农业组织出版物

FAO 是一个致力于减轻全球饥饿、提高粮食安全和实现可持续农业发展的国际组织，其出版物涵盖了有关农业、渔业、林业、畜牧业等方面的报告、手册和指南等。

（二）Annual Activity Report-Agriculture and Rural Development——农业和农村发展年度活力报告

这是由欧盟委员会农业和农村发展总局（Directorate-General for Agriculture and Rural Development，DG AGRI）制作的年度报告，涵盖了该地区农业、渔业、林业、人口和气候等方面的数据和趋势。

（三）The World Agricultural Supply and Demand Estimates（WASDE）——世界农业供需估算

一份由美国农业部世界农业展望委员会（World Agricultural Outlook Board，WAOB）发布的月度报告，提供美国和世界小麦、大米、粗粮、油籽和棉花供应和使用的年度预测。该报告还涵盖了美国糖、肉类、禽蛋和牛奶的供应和使用情况，以及墨西哥的糖供应和使用情况。

（四）Agricultural Policies in OECD Countries：Monitoring and Evaluation——OECD 国家农业政策：监测与评估

这份报告由经合组织（OECD）编写，分析成员国的农业政策，对比不同国家和地区之间的经验和措施效果，并提出了相应的建议和政策推荐。

（五）The State of Food Security and Nutrition in the World——世界粮食安全与营养状况报告

这是由联合国粮食及农业组织、世界卫生组织、联合国儿童基金会、世界粮食计划署和国际货币基金组织等机构联合发布的一份全球性报告，评估了全球范围内粮食安全和营养状况的现状和未来趋势，并提供了解决问题的官方建议方案。

上述出版物覆盖了全球各地农业发展的多个方面，涉及政策法规、市场信息、科技进步、环境保护等多个领域，为全球农业的研究与决策者提供了有用的指导性文献。

四、美国政府农业出版物

美国农业部（USDA）在其网站上提供了大量的政府出版物，以下是几个具有代表性的：

（一）USDA Agricultural Projections to 2030，《USDA 2030 年农业预测》

该出版物呈现了美国农业经济未来 10 年的预测数据和情景分析。

（二）National Agricultural Statistics Service Reports by Year，《国家农业统计服务年报》

这份报告由美国农业部国家农业统计局编写，涵盖了各类农产品的生产、价格、贸易、支出和收入等方面的数据和趋势。

（三）Rural America at a Glance，《美国乡村概览》

一份由美国农业部经济研究局（ERS）发布的年度报告，旨在了解乡村地区的经济、社会和人口状况。

（四）USDA Organic Regulations，《USDA 有机认证规定》

有机食品在美国市场中日益重要，这个规定详细描述了认证有机产品的标准和程序。

（五）Dietary Guidelines for Americans，《美国人膳食指南》

美国农业部与卫生与公共服务部联合编写的一份指南，为美国人民提供了健康膳食和生活方式的建议。

以上出版物仅是众多美国农业领域的政府出版物的冰山一角，这些出版物涵盖了政策法规、生产技术、市场信息、健康指导等多个方面。

 课后练习

与在同一研究领域学习的同学以小组为单位合作，找一份属于政府出版物的文件，并与全班同学分享。

第七节 专利

一、定义

专利是指根据国家法律规定，为了保护发明者的创造成果，使其得到专有权利而由国家授予的一种独占性权利。因此，在一般情况下，专利权人可以阻止他人在未经其许可的情况下制造、使用、销售或进口其专利发明。专利权属于知识产权的一种，专利保护在促进科技创新、增强企业核心竞争力、维护市场秩序等方面起着至关重要的作用。

二、特点

专利的主要特点包括以下几个方面：

（一）独占性

即发明者获得对该发明在一定期限内的排他性使用权。

（二）创造性

专利申请必须符合"新颖性""创造性""实用性"的要求，保证发明具有相对独立性和技术进步性。

（三）有限性

专利权期限在不同国家和地区不尽相同，但通常为 20 年左右，自专利申请日起计算。

（四）地域性

专利权只在授予专利的国家和地区有效。

（五）公开性

专利发明必须公开，以供其他人获取，同时社会大众可以通过公开的专利信息了解相关技术和发明。

（六）可转让性

专利权可以出售或许可给他人使用、生产或销售。

三、专利申请

专利申请是指向国家知识产权局递交一份专利申请书，请求授予所申请的发明或实用新型专利权的行为。一般来说，专利申请程序包括以下几个环节：

（一）提交申请

首先需要填写专利申请书并提交到国家知识产权局或其授权机构。

（二）检索与初步审查

国家知识产权局会对申请文件进行检索与初步审查，以确定是否满足专利保护的基本条件。

（三）公开与实质审查

经过初步审查合格后，申请人可以选择公开申请文件，并进入实质审查阶段。在这个阶段，国家知识产权局会对专利申请的真伪、新颖性、创造性和可行性等方面进行审查，决定是否授予专利权。

（四）授权与维持

如果申请人通过实质审查，将获得专利权授权，然后，申请人需要缴纳一定的年费来维持专利权。

一份完整的专利申请，至少包括申请书、说明书、权利要求书和摘要4个材料，此外还可能包括附图和申请人授权书，各个材料的主要内容如下：

（一）申请书

是一份介绍发明内容、说明技术特点、要求专利保护的文件。需要遵循专利法律法规的相关要求，描述发明内容的优点和实用性，并规范格式。

（二）说明书

是申请书的重要组成部分，必须详细、准确、清晰地阐述发明内容及其最佳实施方式，说明为解决现有技术所存在的问题而采取的技术手段及其效果。

（三）权利要求书

是指根据说明书抽取出来的用于确定专利权利的具体范围和要求，需要精练、明确地阐述发明实质和限制范围。权利要求书是申请书中最重要的部分，因为其限定了申请专利保护的主题。这意味着，一旦授予专利权，专利保护的范围就由权利要求书中所写的内容来决定。

（四）摘要

是发明的简要概述，需要包括技术领域、背景、解决问题的方法、实施方案和效果等，全面准确地介绍了发明的主要特点。

（五）附图（如有必要）

特指化学、物理、机械等技术领域中的图纸或者图片，在说明书中加以引用，用以补充说明发明内容，有助于申请人表达发明内容及其技术特点。

（六）申请人授权书（如有代理）

申请人委托专利代理人代表其向专利局提出专利申请，授权书需要明确申请人和代理人的姓名、地址、联系方式等基本信息，并包括授权范围、专利申请号等相关具体内容。授权书需加盖申请人和代理人的签章或者公章，并需要在法律上获得有效的认证或公证。同时，在撰写授权书时需要注意保护申请人权益和保密要求，明确代理期限和服务内容，并遵守相关法律法规的规定。如果您在申请专利时需要委托专利代理人为您代理申请，建议您选择正规的专利代理机构，并妥善协商好代理协议、代理授权书等相关事宜。

专利申请材料的语言特点，主要包括以下几个方面：

（一）语言精练

专利申请文件通常要求用简练、清晰的语言表达发明内容和技术特点，避免冗长、复杂的描述，使人容易理解。

(二) 技术规范性

专利申请文件应涵盖所有技术特点和相关细节,并符合专利法律法规的要求。同时,应严格按照专利申请的格式和要求撰写。

(三) 全面性和准确性

申请文件中所提供的信息和数据必须全面、准确,并且应该以事实为依据。任何不实的陈述都可能导致专利纠纷或被废止。

(四) 创新性

申请文件需要详细说明发明的创新之处并突出技术创新点,使其与已有的技术区分开来。

(五) 逻辑性强

申请文件应当有秩序地组织、叙述,条理清楚、构思合理,同时各部分之间的关系也应当协调统一。

四、类型

各个国家的专利类型主要包括以下4种:

(一) 实用新型专利(Utility Model Patent)

该专利通常被称为"小型专利"或"创新性专利",更适用于短期和较容易获得保护的技术。该类型的专利授予人们对产品或机械装置等方面的保护。

(二) 发明专利(Invention Patent)

该专利是最广泛使用的专利类型,适用于产品、方法或者其改进所提出的新的技术方案,授权前须实质审查。

(三) 外观设计专利(Design Patent)

该专利适用于某个物品的外观设计方面的保护,例如产品的造型、图案、颜色等。因此,外观专利与实用专利是完全不同的。

(四) 植物专利（Plant Patent）

该专利是指针对新品种植物及其繁殖的知识产权保护。在植物专利中，独特的种植方案被证实为创新性，未经申请者许可的繁殖被认为侵犯该种植物的知识产权。

此外，各个国家的专利还包括商标专利、软件专利等其他类型，这些不同类型的专利分别保护不同范畴内的知识产权。

五、版权

版权（Copyright）和专利（Patent）都是知识产权领域中的重要概念，两者之间有许多相似之处，但也存在一些不同之处：

（一）保护对象不同

版权主要是针对著作、艺术、音乐等智力产品的创作形式，而专利则更多是针对新型发明、实用新型和外观设计等技术性的发明创造。

（二）权利范围不同

版权所涉及的内容较为广泛，包括文学、音乐、艺术、戏剧、电影、软件等。专利的权益范围主要在于向公众公开发明，并被授予独占性使用权或部分共同使用权。

（三）权利产生时间不同

版权的产生是基于作者自行创作时即已经产生了版权，无须类似专利那样进行严格申请审查流程。而与此不同，专利必须在提交申请并受到审查通过后，才能获得其申请的效力。

（四）有效期限不同

目前，美国、英国、日本等国的国内法规定一般版权保护期为作者终生及其死后70年，但我国《著作权法》规定为作者终生及其死后50年。在我国，如果是合作作品，版权保护的期限截止于最后死亡的作者死亡后第50年的12月31日。职务作品的权利的保护期为首次发表后第50年的12月31日。而专利的保护时间通常较短，可根据不同的国家和地区而有所不同。

（五）维权方式不同

版权纠纷多采用民事诉讼方式解决；而在发现专利侵权时，可以将证据固定后直接向相关的法院起诉，也可以选择向管理专利工作的部门进行投诉，后者叫作专利行政执法。与诉讼相比，行政执法的门槛比较低，所以适用范围比较广。此外，并购或重组过程中，还可能涉及不同国家之间的跨境专利诉讼，需要通过国际仲裁等方式解决。

总之，版权和专利在知识产权领域中在保护对象、权益范围、权利产生时间、有效期限及维权方式等方面存在差异。作者需对这些差异有清晰认识，在适当的情况下选择最适合自身权益需求的一种方式来进行申请和管理。

 课后练习

与在同一研究领域学习的同学分组合作。试着找一个专利和一本专著，比较它们在风格、格式、写法等方面的异同。

第八节 标准

一、定义

标准是指提供要求、规范、指南或特性的文件，这些文件可被一致地使用，以确保材料、产品、过程和服务的安全性、可靠性和一致性。

基于科学、工业和消费者的经验，标准也被证明是促进技术进步、专业化和社会生产的科学协调的先决条件。

二、特点

标准是指在一定的范围内，经过共同的研究和理论探讨，由有关方面制定出来、公布并推广应用的一种明确规定性文件，标准具有以下特点：

（一）规范性

标准是针对某一领域或对象而制定的，具有强制性、权威性和规范性。

（二）统一性

标准是通行的、普遍适用的，旨在达成协调、统一的效果，促进全球市场交流与合作。

（三）具体性

标准是一种明确的、切实可行的行动指南，要求在规定范围内做到符合统一规格和技术要求。

（四）可操作性

标准应该能够在实践中得到操作，也就是标准制定时要考虑实际条件和实际可行性。

（五）及时性

标准应该随着时代和科技的发展不断更新和改进，以满足新形势下的需求和挑战。

总之，标准是一种规范性文件，通过制定明确的规定和技术要求，推进特定领域的诸多互动和合作，以达到统一认知、协同行动的目标。同时，标准还可通过推广实施促进经济、科技与社会的发展进程。

三、类型

标准种类繁多，从不同角度来看，标准可以有以下分类：

（一）根据标准适用领域分类

标准可分为行业标准和通用标准。行业标准是特定行业所使用的，只涉及某个行业内的规定。而通用标准（如 SI 国际标准）则是跨越多个行业、领域适用的标准，其覆盖面往往更广泛。

（二）根据标准依据的不同分类

标准可分为基础标准与应用标准。基础标准通常是用于规范测量单位、物理常数、命名等基础性的公认规范；而应用标准则涵盖任何可记录、计算且能够用于实际生产和服务需求的标准。

（三）按照标准的适用范围分类

标准可分为国家标准、地方标准、行业标准、通用标准、企业标准和国际标准等。国家标准适用于全国，地方标准适用于相应区域，在产品质量、安全管理和环境保护等方面起着重要的作用；行业标准主要是针对某个特定行业的技术标准和管理标准；企业标准则是由某一个企业或集团自主制定的标准，通常用于其内部生产和管理；国际标准则是被多个国家广泛认同的标准。

（四）按照标准的形式分类

标准可分为法律标准、行政标准、推荐性标准、指导性标准和强制性标准等。法律标准具有法律效力，是各类标准中最具权威性的标准；行政标准是政府在行政工作中所使用的规范和标准；推荐性标准只起到建议、指导和推行的作用，没有强制性；指导性标准则是针对某一行业或领域的知识普及和技术水平提高等目的而制定的标准；强制性标准是具有强制力的标准。

（五）按照标准的作用分类

标准可分为产品标准、服务标准、管理标准和测试方法标准等。产品标准主要是针对某一种产品特定的技术性要求；服务标准是针对服务行业的技术性和管理性要求；管理标准是针对公司、组织、工厂等内部规章制度和操作流程的标准化要求；测试方法标准则是指确定产品质量的试验、检验、分析标准和技术条件标准等规范标准化方法。

四、ISO 和 ISO 9000 标准

ISO 是国际标准化组织（International Organization for Standardization）的缩写，成立于 1947 年，总部位于瑞士日内瓦，由来自各个国家的专家和行业代表组成，旨在促进全球商贸、技术交流和科研合作等多领域的发展。ISO 制定并发布了众多与经济、环境、质量管理、食品安全、医疗卫生等不同领域相关的标准，目前已有超过 20 000 个

ISO 标准被制定出来。

ISO 9000 标准则是一套以质量管理体系为核心的标准体系，其主要目的是指导企业建立、实施和维护其自身质量体系，并为更好地满足客户需求和提高产品或服务的品质水平提供外部认证。ISO 9000 标准最初发布于 1987 年，目前已经更新到第四版 ISO 9001：2015。这套标准体系包括 ISO 9001、ISO 9002 和 ISO 9003 三个单独的标准，其中 ISO 9001 是最具权威性和适用性的标准，其他两个标准则已经被淘汰。

ISO 9001 标准适用于所有类型和规模的组织，无论其所在的行业和领域，包括制造业、服务业、政府机构、非营利组织等。该标准主要涉及组织的质量管理体系架构、质量目标设定和实施、流程控制、员工素质培训、持续创新改进等方面。实施 ISO 9001 标准可以帮助企业提高产品或服务的品质，增强客户信任并遵循行业最佳实践。

总之，ISO 是国际标准化组织，通过制定和发布各种标准来促进全球商贸、技术交流和科研合作；而 ISO 9000 则是以质量管理体系为核心的一套标准体系，旨在帮助企业建立、实施和维护质量管理体系，提高产品或服务的品质水平，并获得外部认证。

五、国际种子检验标准

《国际种子检验规程》（International Rules for Seed Testing）是由国际种子检验协会（International Seed Testing Association，ISTA）制定、修订和出版发行的种子扦样和检验方法的标准，又称 ISTA 规程，该标准是 1 000 多个物种的发芽条件和检验方法的有用参考指南。

自 1924 年成立以来，ISTA 一直致力于制定、采用和发布种子扦样和检验的标准方法，并推动这些方法在国际上得到统一应用。《国际种子检验规程》已成为全球种子质量评估的权威方法，也是种子质量检测技术研究的重要参考依据。

第一版《国际种子检验规程》于 1931 年出版，随后每 3 年修订一次，自 2001 年以来则每年都有新修订版本。规程的发布方式从最初的纸质版发展为纸质版和电子版，自 2004 年起则仅有电子版可供获取。

ISTA 积极鼓励和推动新物种检验方法的研发和纳入规程中。《国际种子检验规程》的修订是在 ISTA 技术委员会不断研发、采用和评估新方法的基础上进行的。技术委员会每年将拟修订内容方案提交给 ISTA 年度例行大会，然后由各成员国和地区指定代表进行投票表决。表决通过的修订内容将替换原有内容，并编入新版规程，于次年 1 月 1 日正式生效。

现行《国际种子检验规程》（2024 版）主要分为 19 章，每章种子质量参数检验方法的编写格式都非常详细，包括目的、定义、原则、仪器、程序、计算和结果表示、结果报告以及容许差距。其中，第一章规定了 ISTA 证书的使用要求，第二章详细说明了扦样的方法，确保从大量种子批中取样的代表性。其余 17 章则涵盖了种子质量参数的检验方法，具体包括净度分析、其他植物种子数检验、发芽试验、生活力生物化学检验、种子健康检验、物种和品种检验、水分检验、千粒种子重量检验、包衣种子检验、离体胚的活力检验、通过称重重复检验种子、X 射线检验、种子活力检验、种子大小和分级规则、散装集装箱、混合种子分析、转基因种子检验。

 课后练习

与在同一研究领域学习的同学以小组为单位合作。试着找到一个属于标准类别的文档，并与全班分享。

第九节 科学新闻

一、定义

科学新闻是对科学技术领域最近发生事件的真实报道，可以是科技成果及其应用、科学技术政策、科学技术领域的活动，也可以是科技发展趋势预测。通过传播最新的科学研究和创新成果，科学新闻在提高公众科学素养、推动科学进步和人类文明发展中起到了不可替代的作用。

二、特点

科学新闻具有新闻性和科学性，往往有以下几个特点：

（一）通俗易懂

科学新闻采用通俗、易懂的语言，提供一定的背景资料，避免过度专业化术语，以便让更多的读者能够理解和获得知识。

（二）及时性强

科学新闻来源广泛，涉及许多方面的发现和进展，包括各种领域的研究项目、试验和调查结果等，是及时反映最新研究动态的媒介之一。

（三）准确性高

科学新闻需要基于权威的研究机构、期刊或专家提供的信息，保证所发布的内容真实客观可靠。

（四）具有启发性

科学新闻不仅是对新的发现和进展的介绍，而且还可以提供对未来科学发展的启示，促进公众对创新和技术的关注。

三、类型

科学新闻可分为科学新闻报道、科学通讯、科学专题报道和科学新闻评论。

（一）科学新闻故事

科学新闻故事（Science News Stories）是指以科学研究、技术开发、科学政策等为主要内容的新闻文章。这些文章通常从科学家或科学机构得到相关信息，以新闻报道的形式撰写，用通俗易懂的语言向公众传递最新的科学进展和研究成果，并帮助读者理解这些成果对日常生活和社会进步的影响。

与普通新闻报道不同，科学新闻故事更注重科学背景、故事背后的细节和数据，以及代表性的试验结果。旨在为大众呈现多样的科技前沿新闻，观点准确，包括那些突破创新阈值、标志着重大的科技转型过程的事件，例如一项基因编辑技术用于编辑某种农作物，无人机监测农田，或者一个国家推出了一项对气候变化具有重大意义的政策等。

一些著名的科技媒体和网络新闻平台，如 *Science News*、*Nature News*、*Scientific American*、*Wired Science* 等，每天都会发布大量的科技新闻，其中很多都是 Science news stories。

（二）科学通讯

科学通讯（Science Newsletters）是科学界或相关组织、媒体、学术期刊等向专业人士、科技爱好者和订阅者提供的一种定期出版物，包含了有关科学新闻、研究成果、科技进展、政策法规等方面的内容。一些知名的 Science Newsletter 主要包括：

（1）*Science Newsletters*：来自 *Science* 杂志的 Newsletter，每周发布一次，主要包含最新的科学研究发现、科学赛事、会议和其他重要信息。

（2）*Nature Briefing*：来自 *Nature* 杂志的 Newsletter，每天早晨发布，包括最新的科学研究发现、学术竞赛活动等。

（3）*ScienceDirect Newsletter*：由 Elsevier 所运营的一个 Newsletter，主要涵盖各个领域的研究成果，还包括对读者感兴趣的专题报道和特别内容。

（4）*EurekAlert*！：是由美国科学促进会（AAAS）主办的一项全球的互联网新闻服务，提供每日需要掌握的全球学术研究新闻。此外，这个平台还会定期发布学术会议和研讨会的报告，并利用各种媒介形式，如影片和音频，帮助人们更好地了解科学知识和相关研究动态。

通过订阅这些 Newsletter，专业人士和爱好者们能够很快地了解最新的科技新闻、研究成果信息和前沿知识等。同时，也是科学界向公众传递知识和推广科技创新的重要途径之一。

（三）科学专题报道

科学专题报道（Scientific Feature Stories）是一类由科学记者、科学作者或其他专家撰写的深度报道论文，主要为读者提供一个详细的关于某一科学话题或领域的全面性介绍，包括详细的背景资料、各种相关试验的描述、取得的成果及其意义、现在和未来需要面临的挑战等方面，旨在帮助公众全面了解一个科学领域的发展历程。

知名的科学期刊如 *Scientific American*、*Nature* 都有进行科学专题报道的板块。这些平台帮助向大众传达新的科学研究发现，整合当前标志性研究成果，以及预测未来发展趋势。这样的报道能够吸引不同背景的读者，帮助他们理解科学知识，推广科技创新，并为人类未来的生活、工作和学习提供启发和方向。

（四）科学新闻评论

科学新闻评论（Science News Commentaries）是指针对具有重大科学意义的科技新闻、研究成果或科学政策等，由专业人士或著名科学家撰写的分析性评论论文。这类评论通常涵盖更广泛的主题和更深入的分析，不仅讲解一项研究或科学发现的贡献，还试图为公众提供更全面和高层次的解释和观点，并帮助人们意识到某些科学话题可能影响社会和其他领域的发展。

这种科学新闻评论最初在学术期刊、杂志中出现，近年来，跨界合作和多媒体交互的形式得以应用，已经扩展了它们的传播方式和目标受众。一些知名的科学新闻评论平台包括 Science Alert、The Scientist、Nature Opinions 和 Science News 都提供由专家和记者创作的评论论文，评述最新科学研究成果，并对科学界和公共生活中的重大问题表达高层次的看法和分析。这些平台发挥着重要作用来澄清或解析科技信息，为科学与社会之间的沟通架起一座桥梁，并可以帮助读者更全面地认识科学问题的各个方面。

四、世界十大科学新闻

每年，著名的科学网站 www.sciencenews.org 都会评出全球十大科学新闻（如十大科学突破、十大科学发现），追踪科学的最新发展。十大科学新闻排行榜引起了广泛关注。除了这个网站，科技日报社、Nature、环球科学和 CNN（有线新闻网）也评出年度十大科学新闻。虽然这些网站评出的全球十大科学新闻的内容有时并不完全相同，但可以肯定的是，其分享了上一年度科学发展中最重大突破的覆盖面。名单一般在 1 月中旬公布。

 课后练习

作为未来的科学家和工程师，你们需要关注世界上发生的最伟大的发明和创新。现在，试着找出去年世界和中国的十大（农业）科学新闻，并分享给你的班级。

第十节　产品规格书

一、定义

产品规格书（Product Specification）是指一个产品必须符合的相关标准或特定要求。这些规范可以包括诸如尺寸、形状、颜色、材料、性能、测试方法等项目，通常是由国家、行业组织或消费者协会制定并发布的。通过制定产品规格书，可以确保产品具有最高的质量水平，且达到用户所期望的使用标准。

产品规格书是生产制造过程中应该遵循的标准要求，通常由行业组织、国家机构或专业协会制定。产品规格书被看作一种专业文献是因为以下几个原因：

（一）专门化领域

由于产品规格书涉及特定领域中的产品生产和销售，因此需要是质量规格具体性较高地定义了关键技术参数，而与普通消费者接触不够广泛。这些标准和规范的制定都是由相关领域的专家学者或经验丰富的专业人士进行，可以看作是针对某一特定需求设定的专业文章（说明文）。

（二）知识密度大

产品规格书涵盖了许多严格的技术细节和相关条款，其中包括诸如材料性能、工艺流程、质量标准、质量检测等内容，这些都会对产品最终的外形、功能、安全性、可靠性等方面产生重要影响，这些往往需要具有专业背景才可以深入理解。

（三）引用价值高

在生产制造实践中，产品规格书通常被作为参考和指导文献使用，往往会在企业的关键生产流程、技术创新和市场准入方面发挥重要的作用，有时甚至成为法定标准。尤其是在项目实施过程中，通常为了避免可能带来的问题，并最大化满足客户的需求，都需要对产品进行评估并采用切合实际的规范水平要求。

总之，由于产品规格书严谨具体、专业性强、知识密度高，且在产品生产和销售环节起到极其重要的指导作用，因此可以被视为一种专业文献。

二、特点

产品规格书作为一种特殊的专业文献，具有真实性、科学性、逻辑性、实用性等特点。

（一）真实性

产品规格书中的论述是客观的，揭示了产品的本来面目。

（二）科学性

产品规格书作为科学技术应用于实践的结果，代表着科学技术的进步水平，因而具有一定的科学性。

（三）逻辑性

在描述产品时，产品规格书往往从简单到复杂。

（四）实用性

产品规格书强调了产品的实用性，有利于客户的使用。

三、类型

产品规格书按领域可分为：
（一）工业产品规格书
（二）农产品规格书
（三）金融产品规格书
（四）保险产品规格书

四、组成部分

产品规格书的本质是定义和描述产品的设计、功能和质量。因此，产品规格书可能包括设计（设计草案或 3D 设计文件）、颜色（Pantone 或 RAL 颜色）、尺寸和公差（例如 350 ± 5 mm）、重量和公差（例如 400 ± 15g）、材料规格（类型、重量、处理、颜色）、组件（例如型号、品牌、性能要求、功能要求）、认证要求（例如：CE、REACH、FCC）、化学要求（如表面处理、含量）、国际质量标准（如 IP65）、标志（设计文件、尺寸、打印位置、打印类型、颜色）、产品包装（设计、尺寸、布局、颜色、材料）和出口包装（印刷、尺寸、纸箱质量）等。

五、产品规格书和产品说明书的区别

产品规格书（Product Specification）和产品说明书（Product Description）都是介绍产品的相关信息和性能等方面的文献资料，都是为了让消费者更好地了解和使用产品，但产品规格书和产品说明书的内容和用途略有不同。下面是它们的不同点：

（一）产品规格书通常更加技术化和具体，包含产品的各项规格参数、技术指标和

测试方法等，比如产品的尺寸、重量、功率、速度等

（二）产品说明书则更注重产品的使用方法、注意事项、维护保养等方面，比如产品的使用场景、功效、操作指南等

（三）产品规格书更多地面向生产者和技术人员群体，侧重于产品的设计和制造；产品说明书则更多面向消费者，侧重于产品的使用和销售

（四）产品规格书是创造品牌的必须环节，可以借助产品规格书来推广品牌；而产品说明书则更多是关注产品的实际应用，方便消费者使用

 课后练习

与在同一研究领域学习的同学以小组为单位合作。试着找到一些产品规格书，并弄清楚其中包括哪些组件。

第三章

专业英文文献检索与评价

第一节 专业英文文献检索

一、概述

专业英文文献检索是获取、筛选和整合相关文献信息的过程,其重要性和必要性如下:

(一) 获得最新的知识

专业英文文献涵盖了各个领域的最新研究成果和学术进展,通过检索可以及时获取最新的知识和信息。

(二) 提高研究水平

通过阅读并综合分析相关文献,可以拓宽视野、深化理解、提高研究水平和思考能力。

(三) 准确定位问题

通过对相关文献的检索和筛选,可以更加准确地定位研究问题和方向,避免走入盲区或误区。

（四）确认研究现状

通过检索相关文献，可以了解当前研究领域的主要研究方向、前沿领域和热点问题，从而进行更加有针对性的研究。

（五）提高阅读质量

通过综合分析相关文献，可以提高论文阅读的质量和水平，增强论文写作的可信度和说服力。

总之，专业英文文献检索是获取、筛选和整合相关文献信息的过程，具有重要性和必要性。通过检索可以获得最新的知识、提高研究水平、准确定位问题、确认研究现状并提高写作质量，对于学术研究和论文撰写具有不可替代的作用。

二、专业英文文献检索途径

常见的专业英文文献检索途径包括以下几种：

（一）学术搜索引擎

如百度学术、Google Scholar、Bing Academic、Scopus、BASE 等。

（二）学术数据库

如 Web of Science、ScienceDirect、Springer Nature、ProQuest、EBSCO 等。

（三）图书馆目录

如全国图书馆联合目录、美国国会图书馆目录等。

（四）学术社交网络

如 ResearchGate、Academia.edu、Mendeley 等。

（五）学术论坛和博客

如小木虫、Stack Exchange、Quora、Medium、WordPress 等。

（六）预印本网站和资源库

如 arXiv、bioRxiv、SSRN 等。

（七）农业生物学科的网站和资源库

如 PubMed、CAB Abstracts、Agricola 等。

需要注意的是，不同的途径可能对应的文献类型和文献质量不同，因此在进行专业英文文献检索时，需要根据具体情况选择适当的途径。同时，在使用这些途径时，也应该熟练掌握相应的检索工具和检索技巧，以提高检索效率和准确性。

三、专业英文文献检索的一般方法

专业英文文献检索的一般方法包括以下几个步骤：

（一）明确检索目标

确定检索的目的、范围和关键词等信息。

（二）选择检索工具

根据检索目标和需求，选择合适的检索工具，如学术搜索引擎、数据库、图书馆目录等。

（三）制订检索策略

选择合适的检索关键词、使用布尔运算符、通配符、截断符等工具进行检索。

（四）进行检索

按照制定的检索策略进行检索，获取相关文献信息。

（五）筛选文献

对检索结果进行筛选和排序，删除无关文献和重复文献，保留符合要求的文献。

（六）获取全文

对筛选得到的文献，获取其全文。

（七）组织文献信息

对获取到的文献信息进行整理、分类和存储，以便后续查阅和使用。

需要注意的是，在进行专业英文文献检索时，应该尽可能使用多种不同的检索工具和策略，以充分覆盖相关文献，同时避免过度依赖某一种检索工具和策略。此外，对于不同领域的文献检索，其方法和步骤可能会有所不同，需要根据具体情况进行调整和修改。

四、英文专业文献的高级检索方法

以下是专业英文文献的高级检索技巧：

（一）使用布尔运算符

在检索关键词时，可以使用 AND、OR 和 NOT 等布尔运算符来组合关键词。例如，"climate change AND agriculture"将返回同时包含"climate change"和"agriculture"的文献，而"climate change NOT agriculture"则将返回包含"climate change"，但不包括"agriculture"的文献。

（二）利用引号

使用引号将多个单词括起来，以确保搜索结果只包含这些单词的精确匹配（强制检索）。例如，"genetically modified organisms"将返回包含该短语的文献，而不包括仅包含"genetically modified"或"organisms"的文献。

（三）使用通配符

使用通配符（如 *）可以扩展关键词的搜索范围，以便找到与关键词相关的所有变体。例如，"agriculture *"将返回包含"agriculture""agricultural""agriculturist"等词汇的文献。

（四）排除非英语文献

如果想限制搜索结果为英语文献，可以在高级检索界面中设置语言选项，排除其他语言的文献。

（五）搜索特定字段

在高级检索中，可以选择在标题、摘要、关键词或全文中搜索关键词。这有助于缩小搜索范围，提高搜索结果的准确性。

（六）利用限定词

在高级检索中，可以使用限定词（如 author、journal、year 等）来进一步限制搜索范围。例如，在"author"字段中输入某个作者的姓名，将返回该作者发表的所有文献。

总之，利用高级检索技巧可以更加精确地定位所需的文献，提高搜索效率和准确性。需要注意的是，在进行英文专业文献检索时，应该充分掌握各种检索工具的使用方法和技巧，同时也要关注文献来源的权威性和可靠性，避免引入不准确或低质量的文献，对研究造成干扰。

五、利用 Web of Science 检索专业英文文献

基于 Web of Science 检索专业英文文献的方法如下：

（一）登录 Web of Science

在浏览器中输入 Web of Science 网址，登录账号并进入检索界面。

（二）制订检索策略

根据检索目标和需求，选择合适的检索关键词，使用布尔运算符、通配符、截断符等工具进行检索策略的制订。

（三）执行检索

在检索界面输入制定好的检索策略，点击"搜索"按钮进行检索。

（四）筛选文献

对检索结果进行筛选和排序，删除无关文献和重复文献，保留符合要求的文献。

（五）查看文献详细信息

点击文献标题即可查看文献的详细信息，包括作者、摘要、期刊名称、发表日期等。

（六）下载全文

若 Web of Science 库中有全文，则可以直接下载。若没有全文，则需要通过其他方式获取全文。

（七）引用分析

Web of Science 还提供了引用分析功能，用户可以查看某篇文献的被引频次和引用文献列表等信息。

需要注意的是，在使用 Web of Science 进行文献检索时，应该充分利用其高级检索功能，如文献类型、语言、时间范围等过滤条件，以提高检索效率和准确性。此外，在进行文献下载和引用分析时，也应该遵循相关版权和使用规定，不得超越合理使用范围。

六、获取专业文献全文的方法

获取专业英文文献全文的方法有以下几种：

（一）在学校或单位图书馆进行检索

许多大学和研究机构的图书馆都提供了各种在线数据库的访问权限，可以通过这些数据库在网上检索到英文文献，并直接获取全文；若无法获取全文，还可以通过图书馆的文献传递获取。

（二）通过开放获取平台下载

许多期刊和出版社提供了免费开放获取（Open Access）的论文，这些论文可以在其网站上免费下载。例如，PLOS、BioMed Central 等出版社都提供了部分或全部开放获取的期刊。

（三）联系作者

如果需要获取某篇论文的全文，可以尝试通过电子邮件或其他方式联系该论文的作者，请求其提供全文副本。一般来说，对自己的研究内容感兴趣的作者通常会愿意分享他们的论文。

（四）使用文献分享平台

有些学者会将自己的学术论文上传到文献分享平台，例如 ResearchGate、Academia.edu 等，其他学者也可以在这些平台上找到这些论文并进行下载。

（五）购买全文

如果以上方法均无法获取论文的全文，还可以尝试购买该论文的全文。许多商业数据库和在线图书馆都提供了购买文献全文的服务，但需要付费。

需要注意的是，有些文献可能会受到版权保护，不能在互联网上自由传播，因此在获取全文时需要遵守相关法律法规，不要侵犯他人的知识产权。

七、提高英文专业文献检索能力

以下是提高英文专业文献检索能力和水平的方法：

（一）学会使用多种检索工具

熟练掌握不同领域、不同类型的检索工具，如百度学术、Web of Science、Scopus 等。

（二）确定准确的检索关键词

合理选取关键词，包括主题词、同义词、缩写词、通用术语等，扩大检索范围。

（三）制订科学合理的检索策略

采用逻辑运算符、限定符、截断符等工具制订科学合理的检索策略，提高检索效率和准确性。

（四）了解相关领域和前沿动态

关注最新的研究趋势、前沿技术和领域动态，以完善检索策略和提高检索效果。

（五）阅读和分析参考文献

查看已发表文献的参考文献，了解该领域的研究进展和前沿成果，为检索策略调整提供依据。

（六）学习相关专业术语和知识

了解相关专业术语和知识，可以帮助更加深入地理解和分析文献内容，并为检索提供更准确的关键词。

（七）学习文献评价方法

学习文献质量和可信度的评价方法，避免引入低质量或不可靠的文献，影响研究成果的准确性和可靠性。

总之，提高英文专业文献检索能力和水平需要长期积累和实践，同时也需要根据实际需求不断调整和完善检索方法和策略。

八、举例分析

目标：寻找近 5 年内发表的关于小麦（wheat）抗逆性（stress tolerance）育种（breeding）方面的研究文献。

（一）检索词选择

wheat/*Triticum*

stress tolerance/abiotic stress tolerance/drought tolerance/salt tolerance/heat tolerance

breeding/genetic improvement

（二）检索方法

将检索词放入括号，使用布尔运算符连接，构成语句。例如：

（wheat［Title/Abstract］ OR *Triticum*［Title/Abstract］） AND （stress tolerance［Title/Abstract］ OR abiotic stress tolerance［Title/Abstract］ OR drought tolerance［Title/

Abstract] OR salt tolerance [Title/Abstract] OR heat tolerance [Title/Abstract]) AND (breeding [Title/Abstract] OR genetic improvement [Title/Abstract])

在检索工具或数据库中进行全文、标题/摘要等多方面的检索。

（三）时间过滤

将搜索结果按时间排序，选择最近 5 年的文献。

（四）文献类型过滤

选择原创研究、综述、评估等类型的文献，排除会议摘要、书评等非主要研究文献。

（五）结果导出

将符合以上筛选条件的文献导出，并进行进一步的阅读、筛选和分析。

 课后练习

1. 您需要找到有关转基因作物的最新研究文献。您会选择哪个检索工具或数据库，为什么？
2. 您想对某种植物病害进行全面的文献综述，需要涉及跨学科的多个领域，比如植物科学、微生物学、生态学等。您会选择哪个检索工具或数据库，为什么？
3. 寻找近 3 年内发表的关于棉花黄萎病方面的研究文献。

第二节 专业英文文献检索的评价和分析

一、概述

评价和分析专业英文文献检索是指对不同的文献检索工具、数据库或搜索引擎根据特定的指标进行比较和评估，以确定其性能和可靠性。这是研究人员掌握研究领域进展的一项基本技能。

二、专业英文文献检索平台的评价

专业英文文献检索平台的评价可以从以下几个方面进行：

（一）覆盖范围

检索资源的广泛程度和全面性。可以通过比较不同检索工具或数据库之间的期刊、会议论文、书籍、报告等文献覆盖范围和出版商的差异来评价。

（二）检索策略和过滤器

检索工具或数据库提供的多种搜索策略和过滤器，能够帮助用户更精细地控制检索条件和过滤检索结果。可以通过对比不同检索工具或数据库提供的搜索策略和过滤器选项来考察其优劣。

（三）检索效率

检索效率是指在一定时间内，检索系统能够找到满足用户需求的文献数量。评价检索效率的主要指标包括召回率（搜索结果中包含实际所需文献的比例）、精确度（搜索结果中真正相关的文献比例）、搜索速度等。

（四）检索质量

检索质量是指检索系统所提供的文献与用户需求的吻合程度。评价检索质量的主要指标包括覆盖范围（检索系统所涵盖的文献来源）、数据更新速度、检索算法等。

（五）可视化和分析工具

检索工具或数据库提供的可视化和分析工具，可以帮助用户更直观地了解文献信息、引用关系和作者合作等情况。可以通过使用不同检索工具或数据库的可视化和分析工具，体验其功能和效果来进行评价。

（六）用户体验和支持服务

检索工具或数据库提供的用户体验和支持服务，包括网站界面、使用指南、帮助中心、在线客服等。可以通过用户反馈和评价、社交媒体和论坛等途径来了解用户体验和支持服务的好坏。

（七）经济性

经济性是指利用检索系统获取文献所需要的成本和资源消耗。评价经济性的主要指标包括检索系统订阅费用、检索时间成本等。

在对专业英文文献检索的评价和分析中，需要根据具体情况选取合适的指标和方法，以全面、客观地评价检索系统的性能和效果，为后续研究提供参考依据。

三、专业英文文献检索的评价方法

专业英文文献检索的评价方法主要包括以下几个方面：

（一）检索结果的覆盖率和相关性

检查检索结果是否覆盖所需范围和内容，并且是否与研究主题相关。

（二）检索策略的科学性和合理性

评估检索策略的逻辑性、科学性，以及是否充分考虑了关键词、同义词、缩写词、通用术语等因素。

（三）文献来源的可靠性和权威性

评估检索到的文献所在期刊或出版社的可靠性和权威性，如其是否经过同行评审、是否有良好的学术声誉等。

（四）文献质量的可信度和可重复性

评估文献质量的可信度和可重复性，如试验方法和数据分析是否科学合理、结论是否明确、结论是否能被其他研究者验证等。

（五）文献的时间和地域分布

检查检索结果的时间和地域分布，确保涵盖最新的研究进展和全球范围内的相关文献。

（六）数据库和工具的专业性和可靠性

使用专业的文献数据库和检索工具，并评估其可靠性和专业性。

需要注意的是，对于不同研究领域和主题，可能需要考虑的因素也有所不同，因此评价方法需要根据具体情况进行综合分析和判断。

四、专业英文文献检索结果的评价指标

在文献检索结果的评价中，以下是几个常用的性能指标：

（一）查全率（召回率）

查全率指的是检索到所有相关文献数量占全部相关文献数量的比例，其公式为：Recall=TP／（TP+FN），其中 TP 为检索到相关文献数量，FN 为未检索到相关文献数量。当查全率高时，能够检索到更多相关文献。

（二）查准率（精确度）

查准率指的是检索到的相关文献数量占全部检索到的文献数量的比例。其公式为：Precision = TP／（TP + FP），其中 TP 为检索到相关文献数量，FP 为误检文章的数量。当查准率高时，检索到的文献更可能与用户需求相关。

（三）误检率（假阳性率）

指的是与关键词无关但被错误地检索到的文献数量占全部检索到的文献数量的比例。其公式为：False alarm rate = FP／（TN + FP），其中 FP 为误检文献数量，TN 为与关键词无关但未被误检的文献数量。当误检率高时，用户需要耗费更多时间和精力筛选出相关文献。

（四）漏检率（假阴性率）

指的是与关键词相关但未被检索到的文献数量占全部相关文献数量的比例。其公式为：Miss rate = FN／（TP+FN），其中 FN 为未检索到相关文献数量，TP 为检索到相关文献数量。当漏检率高时，存在很多与用户需求相关的文献未被检索到。

上述这些指标都是用来评估检索效果的，但是这些指标之间存在着一定的矛盾关系。比如，提高查准率通常会降低查全率，反之亦然。因此，在实际应用中，我们需要综合考虑各项指标，以达到最佳的检索效果。

五、文献评价方法

文献评价方法是对文献质量和可信度进行评估和判断的方法，主要包括以下几个方面：

（一）论文质量

论文质量是评价英文文献的一个重要标准。优秀的论文应该有足够的数据支持，方法科学合理，并且得出的结论是明确的、可靠的。

（二）研究主题的前沿性和实用性

评价英文文献还需要考虑研究主题的前沿性和实用性。具有较高前沿性和实用性的研究往往更容易引起学术界和社会关注，产生更大的影响力。

（三）出版渠道

出版渠道也是评价英文文献的一项重要指标。权威的期刊或出版社通常会对所发表的论文进行严格的审核和筛选，因此在权威期刊或出版社上发表的论文往往更具有可信度和可靠性。

（四）检查作者资格和背景

了解作者学术资历和背景，如学位、职称、科研经历、学术荣誉等，避免被不具备相关学术背景的人员误导。此外，H 指数是评价学者在自己研究领域内发表高水平论文的数量和被引用次数，是评价其学术影响力的主要指标之一。

（五）查看文献的引用频次和影响因子

检查文献在同行评议中的引用频次及其所在期刊的影响因子这些指标可以反映文献在学术界的重要性和质量。

（六）了解试验设计和数据分析方法

查看文献中试验设计和数据分析方法是否科学合理、逻辑严谨，是否有可能产生误导或错误结论。

（七）利用专业数据库和工具进行评价

利用专业的文献数据库和工具，如 Web of Science、Scopus、PubMed 等，通过检查文献的引用情况、作者信息、期刊来源等方面进行评价。

综上所述，评价英文文献需要从多个方面进行考量，包括论文质量、出版渠道、引用次数、作者背景以及研究主题的前沿性和实用性等。需要注意的是，文献评价方法不是绝对的，需要根据具体情况进行综合判断和分析，同时也要保持开放、客观的态度，避免盲目追求某些指标或评价方法。

六、专业英文文献的评价案例

以下是评价一个关于小麦耐旱候选基因鉴定英文文献检索结果的示例：

（一）研究课题

Identification of candidate genes for drought tolerance in wheat

（二）标题检索式

wheat AND drought AND gene

（三）数据库检索

通过 Web of Science 核心合集检索到 186 个结果（2023-04-24），以下是引用频次靠前的 5 篇论文：

1. Luna CM, Pastori GM, Driscoll S, Groten K, Bernard S & Foyer CH. Drought controls on H_2O_2 accumulation, catalase（CAT）activity and CAT gene expression in wheat. *Journal of Experimental Botany*（一区 TOP 期刊）2005；56：417-423.（被引频次：195）

2. He GH, Xu JY, Wang YX, Liu JM, Li PS, Chen M, Ma YZ & Xu ZS. Drought-responsive WRKY transcription factor genes *TaWRKY*1 and *TaWRKY*33 from wheat confer drought and/or heat resistance in *Arabidopsis*. *BMC Plant Biology*（二区期刊）2016；16.（被引频次：193）

3. Ma DY, Sun DX, Wang CY, Li YG & Guo TC. Expression of flavonoid biosynthesis genes and accumulation of flavonoid in wheat leaves in response to drought stress. *Plant Physiol-*

ogy and Biochemistry（二区期刊）2014；80：60-66.（被引频次：236）

4. Hu W, Huang C, Deng XM, Zhou SY, Chen LH, Li Y, Wang C, Ma ZB, Yuan QQ, Wang Y, Cai R, Liang XY, Yang GX & He GY. *TaASR*1, a transcription factor gene in wheat, confers drought stress tolerance in transgenic tobacco. *Plant Cell and Environment*（一区 TOP 期刊）2013；36：1449-1464.（被引频次：167）

5. Liu ZS, Xin MM, Qin JX, Peng HR, Ni ZF, Yao YY & Sun QX. Temporal transcriptome profiling reveals expression partitioning of homeologous genes contributing to heat and drought acclimation in wheat (*Triticum aestivum* L.). *BMC Plant Biology*（二区期刊）2015；15.（被引频次：216）

（四）筛选和评价

保留论文：Hu et al.,（2013），Ma et al.,（2014）和 Liu et al.,（2015）

论文质量评价：这些论文都采用了现代分子生物学和遗传学技术，如转基因、RNA 测序和基因表达分析等，从不同的角度探究了小麦抗旱机制和相关基因。论文的试验设计和数据分析方法都相对合理和可靠。

出版渠道评价：这些论文都发表在权威的学术期刊上，如 *Plant Cell and Environment*、*Plant Physiology and Biochemistry* 和 *BMC Plant Biology*。这些期刊都有较高的影响因子和排名，而且都经过严格的同行评议流程。

引用次数评价：这 3 篇论文都被其他学者广泛引用，反映了它们在该领域中的重要性和影响力。

 课后练习

1. 根据自己的研究方向或兴趣，设计一个英文文献检索的题目，并列出至少 3 个相关的关键词。

2. 在一个学术数据库中进行检索，找到 10 篇与题目相关的英文文献。注意以下几个方面：
论文的题目和摘要是否与检索关键词有关？
论文发表的时间是否符合需求（如是否过时或太新）？
论文所在期刊的影响因子和排名如何？

3. 对上述 10 篇文献进行筛选和评价，保留其中 3 篇最具代表性和可信度的论文。

注意以下几个方面：

论文质量评价：检查论文的试验设计、数据收集和分析方法是否合理可靠；

出版渠道评价：考虑期刊的权威性和审稿流程等；

引用次数评价：检查论文被引用的次数或被其他学者广泛讨论的情况，以了解该论文在该领域中的重要性和影响力；

4. 比较和总结筛选出来的 3 篇论文的共同点和不同点。注意以下几个方面：

研究主题是否相似？

研究方法是否一致或相似？

结论是否一致，或是否存在差异？

第四章

阅读研究型论文

研究型论文，是对某一学术领域的问题进行研究、表述研究成果的论文。按照研究目的不同，研究型论文可以分为探索性研究、描述性研究、因果性研究 3 类。一篇研究型论文一般由以下几部分组成：

（一）标题和作者归属

（二）摘要

（三）引言

（四）研究方法

（五）研究结果

（六）讨论

（七）结论

（八）致谢

（九）参考文献

以上 9 部分都将在以下单元中具体讨论。

第一节　标题和作者归属

一、标题

标题是论文主题的主要标志，也是文献检索时进行相关判断的主要信息来源。

学术论文的标题通常有以下几个特点：

（1）使用更多的名词或名词短语，或动名词。

（2）简明扼要。

（3）具体。

（4）使用更多的复合词。

（一）标题类型

研究型论文的标题有很多种，常用的有4种：陈述式、描述式、疑问式和复合式。不同类型可以用来吸引不同的读者。

1. 陈述式标题

这类标题是一个声明，陈述了论文的主要发现或结论。

示例：

Metabolite profile of xylem sap in cotton seedlings is changed by K deficiency

2. 描述式标题

这种类型标题描述的论文主题，并没有透露主要的结果或结论。

示例：

Physio-biochemical and proteomic mechanisms of coronatine induced potassium stress tolerance in xylem sap of cotton

3. 疑问式标题

这类标题以问题的形式表明了论文的主题，通常避免了研究的结论。

示例：

Hybridization by grafting: a new perspective?

4. 复合式标题

这类标题由主标题和副标题2部分组成。

示例：

Grafting in cotton: a mechanistic approach for stress tolerance and sustainable development

（二）标题中的提示词

标题中常见提示词如表4.1所示。

表 4.1　标题中的提示词举例

提示词	提示词	提示词
analysis of...	models of...	perspective of...
behavior of...	influence of...on...	practical ways to...
calculation method for...	interpretation of...	problem/question of...
comparison of...	investigation on...	process of...
contrast for...	management/control of...	quality/characteristics of...
design of...	measures to...	reconsideration of...
effect/impact of...on...	mechanism of...	relationship/correlation/association of...
estimation of...	method used for...	research/study on...
evaluation/assessment of...	operation of...	role of...
examination of...	optimization/optimal...	simulation of...
exploration of...	origin of...	synthesis of...

二、作者归属

当您阅读一篇论文时，有必要了解是否有多个作者，如果有多个作者，谁是主要作者。一般来说，第一作者被列在开头，通信作者姓名后通常有星形标记，比如 Zhiyong Zhang *，这意味着如果读者想联系作者的话，这位作者就是要联系的人。而且，您需要找到作者所在的大学或机构，作者的地址或 email 地址，以及个人主页。如果有多个作者，且作者来自不同的大学或机构则会有相应的字母或数字上下对应。

示例：

Physio-biochemical and proteomic mechanisms of coronatine induced potassium stress tolerance in xylem sap of cotton

Xin Zhang[1], Huiyun Xue[1], Aziz Khan[2], Peipei Jia[1], Xiangjun Kong[1], Lijie Li[1], Zhiyong Zhang[1]*

[1] Henan Collaborative Innovation Centre of Modern Biological Breeding, Henan Institute of Science and Technology, Xinxiang 453003, China.

[2] Agricultural College, Guangxi University, Nanning, 530005, China.

* Corresponding author; Zhiyong Zhang, z_zy123@126.com

在上面的例子中，有 7 个作者。第一作者是 Xin Zhang（张新），其名字后面标注了"1"，表示该作者来自下方标有相同字母"1"的机构——中国新乡市的河南科技学院现代生物育种河南省协同创新中心（邮编：453003）。后面的 Huiyun Xue（薛惠云）、Peipei Jia（贾佩佩）、Xiangjun Kong（孔祥军）、Lijie Li（李丽杰）、Zhiyong Zhang（张志勇）名字后同样标有"1"，说明他们都来自与第一作者相同的机构。而且，Zhiyong

Zhang 这位作者被标上了"*",这就意味着他是通信作者。第三作者 Aziz Khan 名字后面标注了"2",表示该作者来自下方标有相同字母"2"的机构——中国南宁市的广西大学农学院(邮编:530005)。

三、阅读技巧

阅读标题时,需要辨别哪些单词或短语是最基本的,哪些是补充性的,可以根据词语的重要性来排序,下划线编号(参见后面示例),并据此区分主要信息和次要信息。

阅读作者归属时,请确定以下信息:有多少作者?每个作者都属于哪个机构?谁是通信作者?这可以帮助您更好地理解研究型论文的作者归属。

四、举例分析

(一) 举例 1

Effect of mechanical stress on cotton growth and development

Zhiyong Zhang*, Xin Zhang, Sufang Wang, Wanwan Xin, Juxiang Tang, Qinglian Wang

School of Life Science and Technology, Henan Institute of Science and Technology, Xinxiang, Henan, China.

分析:

<u>Effect of mechanical stress</u> on <u>cotton growth and development</u>
     1   2

这是一个相对比较简单的描述性标题,从带下划线的词块数字中,您可以看到标题中最重要的信息(一级信息)是"Effects of mechanical stress(机械胁迫的效应)","cotton growth and development(棉花生长和发育)"可视为二级信息。

这篇论文有 6 位作者,属于同一个机构——中国河南省新乡市河南科技学院生命科学学院,第一作者其姓名后标有"*",表明是本文的通信作者。

(二) 举例 2

Effects of rootstocks on cryotolerance and overwintering survivorship of genic male sterile lines in upland cotton (*Gossypium hirsutum* L.)

Xin Zhang[1,2], Zhiyong Zhang[2], Qinglian Wang[2], Peng Chen[1], Guoping Chen[1],

Ruiyang Zhou[1*]

[1] College of Agronomy, Guangxi University, Nanning, Guangxi, China, [2] Cotton Research Institute, Henan Institute of Science and Technology, Xinxiang, Henan, China.

分析：

<u>Effects of rootstocks</u> on <u>cryotolerance</u> and <u>overwintering survivorship</u> of <u>genic male sterile lines</u> in <u>upland cotton (Gossypium hirsutum L.)</u>

（标注数字：1、2、2、3、3、4）

从带下划线的词块数字中，可以看到标题中最重要的信息（一级信息）是"Effects of rootstocks（砧木的效应）"；第二个重要信息是"cryotolerance（耐冷性）"和"overwintering survivorship（越冬存活性）"，它们是介词"on"的宾语，可视为二级信息；三级信息和四级信息分别是"genic male sterile lines"和"upland cotton（Gossypium hirsutum L.）"，包含这两个词块的介词短语"of genic male sterile lines in upland cotton（Gossypium hirsutum L.）"对二级信息进行了限定。

这篇论文有 6 位作者：Xin Zhang（张新）、Zhiyong Zhang（张志勇）、Qinglian Wang（王清连）、Peng Chen（陈鹏）、Guoping Chen（陈国平）、Ruiyang Zhou（周瑞阳），分别属于两个机构，Xin Zhang、Peng Chen、Guoping Chen 和 Ruiyang Zhou 属于第一单位——中国广西壮族自治区南宁市的广西大学农学院；Xin Zhang、Zhiyong Zhang 和 Qinglian Wang 属于第二单位——中国河南省新乡市的河南科技学院棉花研究所；可以看出，第一作者同属于这两个机构或先后在这两个机构从事研究。第 6 作者 Ruiyang Zhou，其姓名后标有"*"，是通信作者。

（三）举例 3

Integration of conventional and advanced molecular tools to track footprints of heterosis in cotton

Zareen Sarfraz[1†], Muhammad Shahid Iqbal[1,12†], Zhaoe Pan[1], Yinhua Jia[1], Shoupu He[1], Qinglian Wang[2], Hongde Qin[3], Jinhai Liu[4], Hui Liu[5], Jun Yang[6], Zhiying Ma[7], Dongyang Xu[8], Jinlong Yang[4], Jinbiao Zhang[9], Wenfang Gong[1], Xiaoli Geng[1], Zhikun Li[7], Zhongmin Cai[4], Xuelin Zhang[10], Xin Zhang[2], Aifen Huang[11], Xianda Yi[3], Guanyin Zhou[4], Lin Li[9], Haiyong Zhu[1], Yujie Qu[1], Baoyin Pang[1], Liru Wang[1], Muhammad Sajid Iqbal[1], Muhammad Jamshed[1], Junling Sun[1*] and Xiongming Du[1*]

[1] State Key Laboratory of Cotton Biology/Institute of Cotton Research, Chinese Academy of Agricultural Sciences (ICR, CAAS), P.O. Box 455000, Anyang, Henan, China.

[2] Henan Institute of Science and Technology, Xinxiang, China.

[3] Cash Crop Institute, Hubei Academy of Agricultural Sciences, Wuhan, China.

[4] Zhongmian Cotton Seed Industry Technology CO., LTD, Zhengzhou, China.

[5] Jing Hua Seed Industry Technologies Inc, Jingzhou, China.

[6] Cotton Research Institute of Jiangxi Province, Jiujiang, China.

[7] Key Laboratory of Crop Germplasm Resources of Hebei, Agricultural University of Hebei, Baoding, China.

[8] Guoxin Rural Technical Service Association, Hebei, China.

[9] Zhongli Company of Shandong, Shandong, China.

[10] Hunan Cotton Research Institute, Changde, China.

[11] Sanyi Seed Industry of Changde in Hunan Inc, Changde, China.

[12] Cotton Research Station, Ayub Agricultural Research Institute, Faisalabad, Pakistan.

分析：

<u>Integration of conventional and advanced molecular tools</u> <u>to track footprints of heterosis in cotton</u>
　　　　　　　　　2　　　　　　　　　　　　　　　1

这是一个描述性标题，从带下划线的词块数字中，可以看到标题中最重要的信息（一级信息）是"to track footprints of heterosis in cotton（追踪棉花杂种优势的痕迹）"；第二个重要信息是"Integration of conventional and advanced molecular tools（整合常规和先进分子工具）"。

这篇论文有多达32位作者，分别属于12个机构，前两位作者后都标注有"†"，说明二者是共同第一作者，意思是对本文的贡献最大且相等（常见的共同第一作者的人数是2人，3人或3人以上是共同第一作者的情况比较少见，有些情况下是不认可共同第一作者的）。最后两位作者其姓名后标有"*"，都是本文的通信作者。

 阅读练习

1. 阅读以下标题

识别它们属于哪种标题类型，并根据重要性对标题中的单词和短语进行排序。

(1) Lack of K-dependent oxidative stress in cotton roots following coronatine-induced ROS accumulation

(2) Xylem sap in cotton contains proteins that contribute to environmental stress response and cell wall development

(3) Major gene identification and quantitative trait locus mapping for yield-related traits in upland cotton (*Gossypium hirsutum* L.)

(4) Plant salinity stress response and nano-enabled plant salt tolerance

(5) Genome-wide dissection of hybridization for fiber quality- and yield-related traits in upland cotton

2. 阅读下列标题和作者单位

确定每篇论文有多少作者，通信作者是谁，以及每个作者属于哪个机构。

(1) Genetic analysis of cryotolerance in cotton during the overwintering period using mixed model of major gene and polygene

ZHANG Xin[1,2], LI Cheng-qi[2], WANG Xi-yuan[1], CHEN Guo-ping[1], ZHANG Jin-bao[2] and ZHOU Rui-yang[1]*

[1] College of Agriculture, Guangxi University, Nanning 530005, P. R. China.

[2] Cotton Research Institute, Henan Institute of Science and Technology, Xinxiang 453003, P. R. China.

(2) GWAS mediated elucidation of heterosis for metric traits in cotton (*Gossypium hirsutum* L.) across multiple environments

Zareen Sarfraz[1†], Muhammad Shahid Iqbal1[,2†], Xiaoli Geng[1], Muhammad Sajid Iqbal[1,2], Mian Faisal Nazir[1], Haris Ahmed[1], Shoupu He[1], Yinhua Jia[1], Zhaoe Pan[1], Gaofei Sun[3], Saghir Ahmad[2], Qinglian Wang[4], Hongde Qin[5], Jinhai Liu[6], Hui Liu[7], Jun Yang[8], Zhiying Ma[9], Dongyong Xu[10], Jinlong Yang[6], Jinbiao Zhang[11], Zhikun Li[9], Zhongmin Cai[6], Xuelin Zhang[12], Xin Zhang[4], Aifen Huang[13], Xianda Yi[5], Guanyin Zhou[6], Lin Li[11], Haiyong Zhu[1], Baoyin Pang[1], Liru Wang[1], Junling Sun[1]* and Xiongming Du[1]*

[1] State Key Laboratory of Cotton Biology/Institute of Cotton Research, Chinese Academy of Agricultural Sciences (ICR, CAAS), Anyang, China, [2] Cotton Research Institute, Ayub Agricultural Research Institute, Multan, Pakistan, [3] Anyang Institute of Technology, Anyang, China, [4] Henan Institute of Science and Technology, Xinxiang, China, [5] Cash Crops Research Institute, Hubei Academy of Agricultural Sciences, Wuhan, China, [6] Zhongmian Seed Tech-

nologies Co., Ltd., Zhengzhou, China,[7] Jing Hua Seed Industry Technologies Inc., Jingzhou, China,[8] Cotton Research Institute of Jiangxi Province, Jiujiang, China,[9] Key Laboratory for Crop Germplasm Resources of Hebei, Agricultural University of Hebei, Baoding, China,[10] Guoxin Rural Technical Service Association, Hebei, China,[11] Zhongli Company of Shandong, Shandong, China,[12] Hunan Cotton Research Institute, Changde, China,[13] Sanyi Seed Industry of Changde in Hunan Inc., Changde, China.

(3) Association mapping analysis of fiber yield and quality traits in upland cotton (*Gossypium hirsutum* L.)

Mulugeta Seyoum Ademe[1], Shoupu He[1], Zhao'e Pan[1], Junling Sun[1], Qinglian Wang[2], Hongde Qin[3], Jinhai Liu[4], Hui Liu[5], Jun Yang[6], Dongyong Xu[8], Jinlong Yang[4], Zhiying Ma[7], Jinbiao Zhang[9], Zhikun Li[7], Zhongmin Cai[4], Xuelin Zhang[10], Xin Zhang[2], Aifen Huang[11], Xianda Yi[3], Guanyin Zhou[4], Lin Li[9], Haiyong Zhu[1], Baoyin Pang[1], Liru Wang[1], Yinhua Jia[1], Xiongming Du[1]

* Yinhua Jia, jiayinhua_0@ sina. com, * Xiongming Du, dujeffrey8848@ hotmail. com

[1] State Key Laboratory of Cotton Biology, Institute of Cotton Research, Chinese Academy of Agricultural Sciences (ICR, CAAS), P. O. Box 455000, Anyang, Henan, China.

[2] Henan Institute of Science and Technology, Xinxiang, China.

[3] Cash Crop Institute, Hubei Academy of Agricultural Sciences, Wuhan, China.

[4] Zhongmian Cotton Seed Industry Technology Co., Ltd, Zhengzhou, China.

[5] Jing Hua Seed Industry Technologies Inc, Jingzhou, China.

[6] Cotton Research Institute of Jiangxi Province, Jiujiang, China.

[7] Key Laboratory of Crop Germplasm Resources of Hebei, Agricultural University of Hebei, Baoding, China.

[8] Guoxin Rural Technical Service Association, Hebei, China.

[9] Zhongli Company of Shandong, Shandong, China.

[10] Hunan Cotton Research Institute, Changde, China.

[11] Sanyi Seed Industry of Changde in Hunan Inc, Changde, China.

第二节　摘要

一、概述

摘要出现在一篇研究型论文的开头。摘要是对整篇学术论文的描述，一般包括背景

(background)、问题（problem）、方法（methods）、结果（results）以及结论（conclusion）等 5 部分内容，并不属于正文。摘要不仅是论文的描述性指南，还传达了论文的范围和讨论的主题。摘要可以帮助读者快速确定论文是否与他们的研究相关。

从语言学角度看，一篇好的摘要具有准确、简洁、具体、独立、客观、自成体系等特点。

摘要通常由 5 个部分组成：

（一）背景（Background）

研究的背景和内容，其理论基础和意义。

（二）问题（Problem）

正在研究的特定研究问题，研究的目的，以及特定的研究目标或假设。

（三）方法（Method）

研究问题所用的方法，做了什么或测量了什么，怎么做的，研究的程度，研究的地点，什么时候进行的。

（四）结果（Results）

重要的数据和主要发现。

（五）结论（Conclusion）

结果的影响和未来的研究方向。

此外，一些研究型论文存在 objectives——目标（在背景和方法之间）和 limitations——局限性（在摘要的最后）等部分。

二、类型

摘要有 4 种常见的类型：信息型摘要、指示型摘要、结构型摘要和图形型摘要。

（一）信息型摘要

信息型摘要是一份论文的缩影，勾勒出整个论文的轮廓，从论文的每个主要部分提炼了关键信息，并提供了关键事实和结论。通常这种类型的摘要对论文的每个主要部分都用一两句话表示。

示例：

Genetic analysis on cryotolerance of cotton during the overwintering period using mixed model of major gene and polygene

The joint analysis of the mixed genetic model of major gene and polygene was conducted to study the inheritance of cryotolerance in cotton during the overwintering period. H077 (*G. hirsutum* L., weak cryotolerance) and H113 (*G. barbadence* L., strong cryotolerance) were used as parents. Cryotolerance of six generation populations including P_1, P_2, F_1, B_1, B_2 and F_2, from each of the two reciprocal crosses H077×H113 and H113×H077 were all investigated. The results showed that cryotolerance in cotton during the overwintering period was accorded with two additive major genes and additive-dominance polygene genetic model. For cross H077×H113, the heritabilities of major genes in B_1, B_2 and F_2 were 83.62%, 76.84% and 90.56%, respectively; and the heritability of polygene can only be detected in B_2, which was 7.76%. For cross H113×H077, the heritabilities of major genes in B_1, B_2 and F_2 were 67.42%, 68.95% and 83.40%, respectively; and the heritability of polygene only be detected in F_2, which was 6.51%. In addition, the whole heritability in F_2 was always higher than that in B_1 and B_2 in each cross. Therefore, for the cryotolerance breeding of perennial cotton, the method of single cross recombination or single backcross should be adopted to transfer major genes, and the selection in F_2 would be more efficient than that in other generations.

（二）指示型摘要

指示型摘要（有时也称为描述型摘要）仅仅包括研究目的、范围和方法等信息。指示型摘要只是为了帮助读者了解研究型论文的一般性质和范围，并没有详细说明研究的过程。

示例：

Effects of development on monoterpene composition of *Hedeoma drummondii*

Samples of leaves, flowers and whole plants were taken from clonal stock of *Hedeoma drummondii* to determine the effect of developmental age on the monoterpene profile. GLC analysis revealed that there are significant differences in the quantity of major monoterpenes in leaves and flowers of different ages and in plants at different flowering stages. The results are discussed in relation to biogenetic pathways and implications for taxonomic work.

(三) 结构型摘要

结构型摘要由几个段落组成,每个段落前面都有一个副标题,类似于最初的研究型论文。

示例:

Integration of conventional and advanced molecular tools to track footprints of heterosis in cotton

Abstract

Background: Heterosis, a multigenic complex trait extrapolated as sum total of many phenotypic features, is widely utilized phenomenon in agricultural crops for about a century. It is mainly focused on establishing vigorous cultivars with the fact that its deployment in crops necessitates the perspective of genomic impressions on prior selection for metric traits. In spite of extensive investigations, the actual mysterious genetic basis of heterosis is yet to unravel. Contemporary crop breeding is aimed at enhanced crop production overcoming former achievements. Leading cotton improvement programs remained handicapped to attain significant accomplishments.

Results: In mentioned context, a comprehensive project was designed involving a large collection of cotton accessions including 284 lines, 5 testers along with their respective F1 hybrids derived from Line × Tester mating design were evaluated under 10 diverse environments. Heterosis, GCA and SCA were estimated from morphological and fiber quality traits by L × T analysis. For the exploration of elite marker alleles related to heterosis and to provide the material carrying such multiple alleles the mentioned three dependent variables along with trait phenotype values were executed for association study aided by microsatellites in mixed linear model based on population structure and linkage disequilibrium analysis. Highly significant 46 microsatellites were discovered in association with the fiber and yield related traits under study. It was observed that two-thirds of the highly significant associated microsatellites related to fiber quality were distributed on D sub-genome, including some with pleiotropic effect. Newly discovered 32hQTLs related to fiber quality traits are one of prominent findings from current study. A set of 96 exclusively favorable alleles were discovered and C tester (A971Bt) posited a major contributor of these alleles primarily associated with fiber quality.

Conclusions: Hence, to uncover hidden facts lying within heterosis phenomenon, discovery of additional hQTLs is required to improve fibre quality. To grab prominent improvement in influenced fiber quality and yield traits, we suggest the A971 Bt cotton cultivar as fundamental element in advance breeding programs as a parent of choice.

(四) 图形型摘要

图形型摘要是对论文主要研究结果的图形化和视觉化总结。图形型摘要可以是论文的结论性图片，也可以是专门为此目的而设计的插图。图形型摘要抓住了论文的内容，让读者一目了然。

示例（图 4.1）：

Water use of intercropped species: Maize-soybean, soybean-wheat and wheat-maize

A graphical abstract of the effects of different intercropping systems on soil water utilization, plant physiological activity and productivity. Note: The light blue to dark blue bars indicates the amount of water stored in the soil, LER is land equivalent ratio, WUE is water use efficiency, IGR is instantaneous growth rate, Pn is photosynthetic rate, Chll is chlorophyll content. Compensation refers to the process of resource acquisition through temporal and spatial niche differences between crops. Competition refers to the crops are competing resource to against each other.

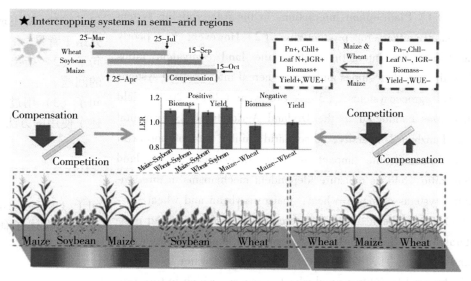

图 4.1　不同间作系统对土壤水分利用、植物生理活性和生产力影响的图解摘要

三、阅读技巧

在阅读一篇研究型论文的摘要时，需要回答下列问题：

（一）为什么要进行这项研究

（二）做了什么

（三）是如何做到的

（四）发现了什么

（五）结论是什么

在试图找到这些问题的答案时，你实际上是在识别一个摘要的5个部分：背景、问题、方法、结果和结论。摘要往往不包含所有的5个部分，有时省略背景或结论。

通常，背景用现在时态，研究问题用现在时态或过去时态，方法和结果用过去时态，结论多用现在时态和将来时态。

四、举例分析

（一）举例1

Water use of intercropped species: Maize-soybean, soybean-wheat and wheat-maize

Abstract: (1) Plant-plant interactions in the intercropping system can significantly affect crop productivity. (2) However, it is poorly understood how the interactions affect the land equivalent ratios (LER) in the cereal-legumes and cereal-cereal intercropping systems in semi-arid agroecosystems. (3) A two-year (2019-2020) field experiment was conducted in the Dryland Agricultural Experimental Station of Lanzhou University, a semiarid rainfed site of northwest China, to quantify the impact of crop diversification on land equivalent ratio, and its moisture-dependent mechanisms in three intercropping systems of maize-wheat, maize-soybean and wheat-soybean. (4) The soybean-involved intercropping systems showed positive interactions, which substantially promoted crop productivity of maize and wheat by 18.1%~20.9%. (5) The soil water in soybean strips can be used by intercropped wheat or maize, which in turn promoted soil water storage (SWS) in maize or wheat strips by 0.6% and 11.0% respectively, during the co-growth period. (6) This further improved the photosynthetic rate (Pn), instantaneous growth rate (IGR), and water use efficiency (WUE) for each species and thereafter elevated the land equivalence ratio (LER>1). (7) However,

句子（1）使用现在时介绍了背景。

句子（2）指出了存在的问题。

句子（3）用过去时描述了使用的方法。

句子（4）~（12）用过去时陈述了研究结果。

in the wheat-maize intercropping system, interspecific competition was dominated with wheat as dominant species and maize as inferior one respectively. (8) The competition plundered available water of maize strips (SWS decreased by 10.5%), and thus decreased its Pn and WUE by 12.7% and 20.0% ($P<0.05$), respectively. (9) Importantly, maize yield and LER were not improved via the compensation effect during the post-harvest period of wheat. (10) While soil water was to some extent restored, the Pn and IGR of intercropped maize were still lower than those of monoculture maize. (11) This trend resulted in maize yield loss during the reproductive period, and ultimately lower LER. (12) For the first time, we found a rarely reported phenomenon, i.e. negative relationships between crop diversity and land equivalent ratios in semiarid agroecosystem, since the yield loss caused by competition was not compensated during the co-growth period. (13) Therefore, a positive relationship between crop diversification and land equivalent ratio required rational crop species configuration, in terms of the tradeoff between crop diversity and its productivity at agricultural landscape scale.

句子（13）总结了研究结论。

（二）举例 2

Physio-biochemical and proteomic mechanisms of coronatine induced potassium stress tolerance in xylem sap of cotton

Abstract：(1) Potassium (K) is a major plant nutrient, and its deficiency can limit plant growth and development. (2) Coronatine (COR) could increase cotton seedling tolerance to K deficiency, which was hypothesized to be in relation to improving the physiological and proteomic profile of xylem sap. (3) To test this hypothesis, cotton seedlings growth, physio-biochemical and proteomic profile of xylem sap treated with (0 and 10 nM COR) under deficient K solution (0.05 mM KCl) were explored. (4) Compared with control, COR treatment significantly increased lateral root number and root diameter and decreased contents of malondialdehyde (MDA), some cations like potassium and calcium in xylem sap and its volume. (5) These morphological and physiological presentations were well evidenced by differentially expressed proteins (DEPs) in the xylem sap. (6) For instance, increasing

句子（1）使用现在时介绍了背景。
句子（2）提出了一个假设。
句子（3）用过去时描述了使用的方法。
句子（4）~（6）用过去时陈述了研究结果。

the average root diameter and several lateral roots were related to up‑regulation of cobalamin‑independent methionine synthase family proteins, auxin‑responsive protein, and cell wall remodelling proteins such as dirigent‑like protein, laccase, and the pectin lyase; lessening xylem sap volume and some cations contents was in connections with down‑regulation of uclacyanin 1 and up‑regulation of calmodulin‑domain protein kinase 7; the MDA content reducing was associated with many PCD‑related proteins' down‑regulation or loss. (7) Furthermore, COR potentially weakened plant defense, owing to lessening or disappearing lipid‑transfer proteins, signaling proteins, and other proteins positively involved in plant defense.

句子（7）总结了研究结论。

 阅读练习

1. 阅读下面的摘要，识别摘要的 5 个部分：背景、问题、方法、结果和结论

Quantitative proteomics‑based analysis reveals molecular mechanisms of chilling tolerance in grafted cotton seedlings

Abstract: (1) Proteome analysis of grafted cotton exposed to low‑temperature stress can provide insights into the molecular mechanistic of chilling tolerance in plants. (2) In this study, grafted and non‑grafted cotton plants were exposed to chilling stress (10℃/5℃) for 7 d. (3) After the stress, rootstock and scion samples were labeled by 8‑plex iTRAQ (isobaric Tags for Relative and Absolute Quantification), followed by two‑dimensional liquid chromatography separation and tandem mass spectrometry identification. (4) In total, 68 differential proteins were identified that were induced by low‑temperature stress and grafting, and these proteins regulate physiological functioning. (5) Under low‑temperature stress, in the cotton seedlings, the proteins responded to the MAPK signaling pathway and calcium signaling pathway enhanced, the metabolisms of carbohydrate, lipid, nucleotide, and amino acid had a tendency to intensify, the proteins related to protein folding and degradation were activated, along with the system of antioxidant enzymes to offset cellular oxidative damage. (6) In contrast, chilling stress reduced oxidative phosphorylation, photosynthesis, and carbon fixation. (7) These data indicated that the physiological changes in cotton seedlings comprise a complex biological process, and the ability of plants to resist this stress can be improved after grafting onto a vigorous rootstock, although this was not obvious in the young plants. (8) Further studies of low-temperature stress and/or graft-related differences in proteins could lead to the identification of new genes associated with chilling tolerance in plants. (9) These data provide the basis for further studies on the molecular mechanism of chilling tolerance and the relationship of grafting and chilling tolerance in cotton.

2. 根据摘要中的部分及其顺序，将表 4.2 中的句子按正确的顺序排列成摘要

表 4.2 摘要组成及其顺序

1	It grows in Yemen and Southern Arabia as well as in certain East African countries such as Ethiopia, Somalia, Djibouti and Kenya. (Background)
2	Khat leaves were collected and bought from Yemen market. The samples were dried in an oven at 50~52℃. The samples were stored in a cool, dark and dry place. The dried sample (200 g) was ground first to pass a 2 mm screen. (Materials)
3	Khat is an evergreen shrub belonging to the Celastraceae family. (Background)
4	It can be concluded that khat contain high amounts of tannins acid, ascorbic acid and fluoride was not found in khat. (Conclusion)
5	The aim of this study was to quantify tannins acid, ascorbic acid and Fluoride from khat extract. (Purpose)
6	Then, tannins acid was detected using Proanthocyanidins technique. Redox titration (Iodometric titration technique) method was used to determine ascorbic acid in khat extracts. Fluoride in khat extracts was detected using a method based on APHA Standard 4500-F-D. (Methods)
7	The Average percentage of tannins acid from khat extracts in dry matter was 25.4%. The average percentage of ascorbic acid extracted from khat in sample was 25.72%. Fluoride substance was not determined in khat. (Results)

第三节 引言

一、概述

Introduction 是学术论文的开头，属于正文部分。"引"字既表示引出文献综述部分，又起到引起读者阅读兴趣的作用。引言解释了研究的理由，并清楚地描述了本研究的主要目的。引言将研究工作置于一个理论背景下，使读者能够达到理解和欣赏研究的目的。一般由以下 4 个语步组成：

（一）确定研究范围

本文从以下几个方面概括一篇论文的主题：

（1）研究在大背景下的意义，包括研究值得做的原因，特别是研究的实践原因和理论原因。

（2）对本课题的概括，包括关键术语、概念特征和性质的一般定义、优缺点、历史发展以及材料或技术的现状。简单地说，这一举动是为了告知读者背景知识。

（二）回顾以往相关研究

这一举措通常是对以往相关研究的总结，包括一个简要的文献回顾（以前和现在在这方面的发展情况），以证明本研究的必要性。

（三）确定一个研究空白

这一举措从指出差距和提出问题开始，为目前的研究做准备。在这一举措中，论文要处理的更广泛主题的特定领域被描述为以下信息：

（1）通过某种方式扩展以前的知识，指出以前的研究中的空白。

（2）研究提出的问题，提出具体的假设并说明理由。

（四）占据这个研究空白

本研究将完成的部分工作包括以下几个方面：

（1）研究的目的，这正是本研究要解决的问题。

（2）本文的总体结构，可以作为试验设计的一个非常简单的描述，以及将如何实现设定的目标。

（3）研究的主要结果，可能是研究完成时所完成的。

二、阅读技巧

在阅读一篇研究型论文的引言部分时，您需要运用以下技巧作为辅助：（1）记住引言部分的4个语步；（2）借助句法和词汇线索识别这4个语步。

（一）确定研究范围

关于研究在广泛背景下的意义，线索可能是谓语时态；指示当前情况的副词或状语短语；诸如 significant 、necessary 、important 等词语。例如：

Coronatine（COR）, a plant phytotoxin similar in structure to jasmonate（Zhang et al., 2021; Zhou et al., 2015）, plays an <u>important</u> role in regulating plant growth, inhibiting senescence, promoting cell differentiation, increasing chlorophyll content, and resisting low K（LK）stress（Shen et al., 2018; Xie et al., 2015）.

Under LK conditions, COR <u>can improve</u> the metabolic activity of cotton roots by increasing their viability as measured by reducing colorless tribenzotriazole chloride（TTC）to red triphenyltoluene（TF）and thus <u>promote</u> the production of cotton lateral roots（Zhang et al., 2015; Zhang et al., 2009）.

关于主题的概括，线索可能是：谓词的现在时；定义、属性、应用等的描述。例如：

Low-temperature stress, including chilling and freezing injury, is a global problem for many field crops. Since 1980, crops on approximately 3.3 million hectares are annually affected by low temperatures in China, accounting for 7.4% of the country's cropping area and up to 10% yield losses（Yan, 2017）.

Cotton（*Gossypium* spp.）originates from the tropics and subtropics and is a cold-sensitive plant（Xu et al., 2017）. The optimum temperature for cotton growth and development is 22~32℃. When the daily temperature falls below 15℃, it affects the growth and development of cotton through chilling injury（Cao et al., 2021）. Low temperature disrupts normal plant growth and developmental processes, reducing overall lint yield and quality（Wang et al., 2014; Wang et al., 2016）.

By comparing cotton cropping systems in Xinjiang, it was found that expansion of cotton-growing areas in the higher altitudes is restricted by low-temperature（Huang and Wang, 1999）.

（二）回顾以往的相关研究

回顾以往的相关研究，线索可能是：谓词的完成时、过去时或现在时；作为句子主体的研究者姓名等。例如：

Several studies have explored and described the defensive processes of plants against biotic（Ali et al., 2014; Delaunois et al., 2014; Yang et al., 2015）and abiotic（Guerra-guimarães et al., 2014; Song et al., 2011; Zhang et al., 2016）stresses by analyzing the apoplast proteomic profile.

Some references reported that the jasmonate receptor COI1（COR insensitive 1）with a high affinity for COR, could activate outward K channel in leaf guard cells and consequent stomatal closure (Munemasa et al., 2007) and that COI1 mediated transcriptional responses of *Arabidopsis thaliana* to external K supply, and *coi*1 mutant presented no significant changes in K content in comparison with wild type (Armengaud et al., 2010).

(三) 确定一个研究空白

研究空白的提示词，线索可能是：谓词的现在时态；诸如 unclear、disclosed、unravel、unsolved、unexplored、shortcoming、drawback 等词。例如：

In cotton, several proteins in the xylem sap are related to its response to environmental stress (Zhang et al., 2016; Zhang et al., 2015), yet the differential proteome of xylem sap induced by COR, and especially its role in root-shoot crosstalk, remain unclear.

However, the molecular mechanism of COR-induced changes in cotton root system under limited K has not yet been disclosed.

Earlier reports unravel that additive and dominance effects laid the foundation of genetics related to heterosis for cotton yield.

为了提出研究问题，作者可能会陈述特定的假设或目标，或一个发现的延伸，并描述选择这些研究问题的原因。例如：

However, to the best of our knowledge, no previous study has investigated the response mechanism of cotton to K deficiency by analyzing the metabolic components of xylem sap.

However, the mixed inheritance of cryotolerance in cotton has not been reported yet.

(四) 占据这个研究空白

研究的目的，线索可能是：谓语的现在或将来时态；诸如 sought to、aim、explore、objective、purpose 等词。例如：

This study sought to quantitatively and qualitatively analyze COR-induced changes in xylem sap proteins and key genes conferring K acquisition in cotton under K-deficiency and to explore the mechanism of COR-induced xylem sap proteins for changing cotton's tolerance ability.

In this study, we used iTRAQ (isobaric Tags for Relative and Absolute Quantification), a high-throughput protein quantification technology to analyze the differential expression of proteomes in the scion and rootstock of grafted cotton seedlings under low-temperature stress, with an aim to understand the molecular mechanism of induced chilling tolerance.

The main objective of the present study was to explore QTLs underlying fiber quality traits in diverse Upland cotton genotypes using SSR markers.

在引言中描述学位论文/研究型论文的整体结构，线索可能是：谓语的现在时或将来时；诸如序数数字 first、second 等词。例如：

The objectives of this study are designed as：(1) to evaluate the differentiate performance of crop productivity in monoculture and intercropping conditions (maize-soybean, wheat-soybean and maize-wheat), (2) to demonstrate the dynamic competitive trajectories of crop growth and development during the co-growth period, (3) to explore the interspecific relationships in different intercropping systems and its association with soil water availability, and (4) to reveal how crop interactions affect the land equivalent ratio via the compensation effect.

三、举例分析

（一）举例 1

(1) Coronatine (COR), a plant phytotoxin similar in structure to jasmonate (Zhang et al., 2021; Zhou et al., 2015), plays an important role in regulating plant growth, inhibiting senescence, promoting cell differentiation, increasing chlorophyll content, and resisting low K (LK) stress (Shen et al., 2018; Xie et al., 2015). (2) Under LK conditions, COR can improve the metabolic activity of cotton roots by increasing their viability as measured by reducing colorless tribenzotriazole chloride (TTC) to red triphenyltoluene (TF) and thus promote the production of cotton lateral roots (Zhang et al., 2015; Zhang et al., 2009). (3) However, the molecular mechanism of COR-induced changes in cotton root system under limited K has not yet been disclosed.

(4) Potassium (K), among the major mineral element essential for plant growth, is the most abundant inorganic cation in higher plants (Khalid, 2013). (5) The K content of plants is generally 1% to 5% of their dry matter weight, accounting for 50% of a plant ash weight (Leigh and Jones, 1984). (6) The role of K in plants consists of two functions: biophysical and biochemical ones (Binepal et al., 2016). (7) The former includes regulating expansion pressure and osmotic potential, while the latter includes promoting enzyme activity, photosynthesis, assimilation transport, protein synthesis, and stress resistance (Jaiswal et al., 2016). (8) Cotton has

通过阅读，句子 (1) ~ (14) 找到语步1，使用的是现在时。其中，句子 (1) 和 (2) 描述了背景——冠菌素可以提高植物耐低钾的能力及其证明方法。句子 (3) 指出了本研究的意义：冠菌素在棉花低钾胁迫下诱导根系的分子机制尚不明确，因此需要深入研究。

句子(4) ~ (10)描述了钾元素及其在棉花生长中的重要性。

a relatively long growth period, large biomass, and relatively high K content in their boll (Wu et al., 2016). (9) Therefore, K deficiency can directly reduce the lint yield and fiber quality of cotton by suppressing reproductive structure formation (Khan et al., 2017). (10) Low K availability in soil adversely affects cotton root growth, photosynthetic capacity (Zia-ul-Hassan and Arshad, 2010), reduces leaf area increases leaf senescence resulting in substantial reductions in boll weight (Tariq et al., 2018).

(11) The apoplast space harboring xylem sap has garnered special attention because it is the first barrier in plants to biotic and abiotic stresses. (12) It can transmit signals to the protoplasts to defense against stress conditions (Delaunois et al., 2013). (13) Xylem sap contains carbohydrates, hormones, and proteins, which move through the apoplast pathway (Rodríguez-Celma et al., 2016); i.e., in the cell walls and interstitial spaces of adjacent cells. (14) Several studies have explored and described the defensive processes of plants against biotic (Ali et al., 2014; Delaunois et al., 2014; Yang et al., 2015) and abiotic (Guerraguimarães et al., 2014; Song et al., 2011; Zhang et al., 2016) stresses by analyzing the apoplast proteomic profile.

句子（11）~（14）回顾了根系木质部汁液中的成分，特别是蛋白质，可以反映棉花在逆境胁迫下的生理响应。

(15) COR can increase plants' resistance to both biotic and abiotic stressors under low K (Xie et al., 2015; Zhang et al., 2015), but evidence for its linkage to proteome dynamics in the apoplast is lacking. (16) A better understanding of how this increase occurs and its mechanisms is of great significance for improving plant resistance under a K-deficient status. (17) Some references reported that the jasmonate receptor COI1 (COR insensitive 1) with a high affinity for COR, could activate outward K channel in leaf guard cells and consequent stomatal closure (Munemasa et al., 2007) and that COI1 mediated transcriptional responses of Arabidopsis thaliana to external K supply, and coi1 mutant presented no significant changes in K content in comparison with wild type (Armengaud et al., 2010). (18) In cotton, several proteins in the xylem sap are related to its response to environmental stress (Zhang et al., 2016; Zhang et al., 2015), yet the differential proteome of xylem sap induced by COR, and especially its role in root-shoot crosstalk, remain unclear.

句子（15）~（18）确定了一个研究空白。

(19) This study sought to quantitatively and qualitatively analyze COR-induced changes in xylem sap proteins and key genes conferring K acquisition in cotton under K-deficiency and to explore the mechanism of COR-induced xylem sap proteins for changing cotton's tolerance ability.

句子（19）占据了这个研究空白。

（二）举例2

(1) Low-temperature stress, including chilling and freezing injury, is a global problem for many field crops. Since 1980, crops on approximately 3.3 million hectares are annually affected by low temperatures in China, accounting for 7.4% of the country's cropping area and up to 10% yield losses (Yan, 2017). (2) Cotton (*Gossypium* spp.) originates from the tropics and subtropics and is a cold-sensitive plant (Xu et al., 2017). (3) The optimum temperature for cotton growth and development is 22~32℃. (4) When the daily temperature falls below 15℃, it affects the growth and development of cotton through chilling injury (Cao et al., 2021). (5) Low temperature disrupts normal plant growth and developmental processes, reducing overall lint yield and quality (Wang et al., 2014; Wang et al., 2016). (6) By comparing cotton cropping systems in Xinjiang, it was found that expansion of cotton-growing areas in the higher altitudes is restricted by low-temperature (Huang and Wang, 1999).

(7) Low temperature is a potent abiotic stress and adversely affects seedling germination and establishment, resulting in inflicting a substantial yield penalty on many important field crops. (8) Cotton is extremely sensitive to low-temperature stress, particularly during vegetative development and reproduction (Zhao et al., 2021). (9) Cold stress thermodynamically depresses the kinetics of many physiological and biochemical processes, i.e., delays germination, seedling vigor reduces starch metabolism and respiration rate in plants, leading to substantial yield losses (Sha et al., 2021).

(10) Under low-temperature stress, the cell membrane is the earliest to respond, i.e., increased cell permeability and cell membrane fluidity (Yang et al., 2020). (11) Low temperature also affects cell physiology, imbalances DNA stability and RNA secondary structures, impairs protein biosynthesis, enzyme reactivity, and photosynthetic performance (Leila and Reza, 2010; Wang et al., 2017; Zhang et al., 2021). (12) Proteomic approaches are powerful tools to study plant responses to environmental stresses.

通过阅读，句子（1）~（9）找到语步1确定研究范围，使用的是现在时。

语步2：句子（10）~（17）回顾了低温对细胞生理的影响，特别是对蛋白质组的影响。

(13) Particularly, proteomics techniques combined with mass spectrometry (MS) can detect translational and post-translational regulation of different proteins (Yan et al., 2006; Wei et al., 2021). (14) Plants respond to environmental stress by modulating the expression or synthesis of proteins (Sarhadi et al., 2010; Janmohammadi et al., 2015; Xu et al., 2018). (15) Thus, changes in plant proteome are a response to abiotic stresses. (16) Proteomics techniques have widely been used in several crops for understanding their responses to abiotic stresses (Wang, 2016; Wang et al., 2021), i.e., low-temperature stress tolerance in corn and sunflower crops (Balbuena et al., 2011; Kołodziejczyk et al., 2016). (17) Physiological and biochemical changes in cotton under low-temperature stress are complex, and proteins usually act together in the context of cellular networks rather than displaying their functions in an isolated manner (Bian et al., 2015).

(18) Currently, cotton grafting is primarily performed for improving crop growth and yield under stressful environments (Zhang et al., 2022), including low-temperature stress (Liu et al., 2004; Zhou et al., 2018; Wu, 2021). (19) Our previous research (Zhang et al., 2012) indicated that chilling tolerance in annual cotton could be achieved by grafting them onto chilling tolerant perennial cotton species. This chilling tolerance in the scion of grafted cotton was positively correlated with total soluble protein contents (Zhang et al., 2013).

语步3：句子（18）~（19）确定了一个研究空白。

(20) In this study, we used iTRAQ (isobaric Tags for Relative and Absolute Quantification), a high-throughput protein quantification technology to analyze the differential expression of proteomes in the scion and rootstock of grafted cotton seedlings under low-temperature stress, with an aim to understand the molecular mechanism of induced chilling tolerance.

语步4：句子（20）占据了这个研究空白。

 阅读练习

阅读以下研究型论文的引言，并根据本单元中提到的4个语步进行分析。把句子的序号放在表4.3右栏。

(1) Environmental biotic and abiotic stresses affect plant metabolism, growth and productivity. (2) During stress, electrons at high energy-states are transferred to molecular oxygen (O_2), resulting in reactive oxygen species (ROSs), including hydrogen peroxide (H_2O_2) and superoxide (O_2^-) and hydroxyl ($\cdot OH$) radicals, which are toxic to plants at high concentrations (Ashraf, 2009; Mittler, 2002). (3) Among the peroxides, H_2O_2 is the most widely spread molecule involving in a broad range of physiological processes, such as growth, development and senescence (Triantaphylides et al., 2008). (4) For protecting the oxidative stress induced by biotic and abiotic factors, plants have evolved many resistance mechanisms to prevent ROS toxicity, including antioxidant systems that are comprised of different non-enzymatic antioxidant compounds, including carotenoids, glutathione, ascorbic acid and vitamin E, and antioxidant enzymes, for example superoxide dismutase (SOD; EC1.15.1.1), catalase (CAT; EC1.11.1.6), guaiacol peroxidase (GPX; EC1.11.1.7), ascorbic acid peroxidase (APX; EC1.11.1.11) and glutathione reductase (GR; EC1.6.4.2). (5) SOD is responsible for the detoxification of O_2^- into H_2O_2 and O_2 (Scandalios, 1993), and CAT and a variety of peroxidases, including GPX and APX, collaborate with GR in the Halliwell-Asada cycle to metabolise H_2O_2 (Alscher et al., 1997; Asada, 1984). (6) Potassium (K) is an important macronutrient affecting plant growth and development, yield, and resistance to environmental stresses (Cakmak, 2005; Hafsi, 2004). (7) K deficiency has been shown to increase H_2O_2 production and up-regulate the peroxidase gene expression in *Arabidopsis* thaliana and tomato roots (Armengaud, 2004; Shin and Schachtman, 2004; Hernandez, 2012). (8) Therefore, antioxidant enzymes were up-regulated under oxidative stress in response to potassium deficiency.

(9) Coronatine (Gifford et al., 2008), which is a novel type of plant growth regulator that possesses similarities in structure and properties to jasmonate, has been shown to significantly stimulate the production of secondary metabolites compared with jasmonate (Uppalapati et al., 2005). (10) Several studies have suggested the potential application of COR in the regulation of stress resistance in crops, i.e., increasing drought resistance in rice and upland rice (Ai et al., 2008) and winter wheat (Li et al., 2010) and improving salt tolerance in cotton (Xie et al., 2008).

(11) Cotton is an important economic crop worldwide, and these plants undergo premature senescence in response to K deficiency (Zhang et al., 2007). (12) Previous experiments have shown that COR enhances lateral root formation, K uptake and seedling growth in

cotton plants (Zhang et al., 2009) and root vitality regardless of the presence of K-sufficient or K-deficient conditions. (13) However, the positive role of COR in root vitality under non-stress or under K-deficient conditions in ROS and antioxidant non-enzymatic and enzymatic systems remains unclear. (14) The objective of this work was to examine the effect of COR on the biochemical defence mechanism of cotton under K deficiency stress.

表4.3 研究型论文引言4个语步练习

引言"4大步"	与该语步相关的句子序号
语步1——确定研究范围	
语步2——回顾以往相关研究	
语步3——确定一个研究空白	
语步4——占据这个研究空白	

第四节 研究方法

一、概述

研究型论文的方法部分描述了研究是如何进行的，主要是记录所用的专门材料和操作程序，以便他人可以在其研究中使用部分或所有的方法，或判断该研究工作的科学价值。研究方法部分不是一步一步地描述研究中的每一件事，也不是一套指令，只需要包括过程和关键信息。

一般来说，这一部分包括材料、仪器、对象、程序、数据收集和数据分析。

在方法部分，读者将阅读研究中使用的各种材料的说明，其中可能包括试剂（用于进行试验的化学品）及其数量来源和制备方法；生物制剂及其来源、属、年龄、性别和生理状态等；设备（研究中使用的技术设施）及其技术规格，如生产、模型、来源等。

工具方面，读者可以找到做试验或收集数据所用的工具的信息，如设备、试卷、问卷、访谈提纲等。

研究中的受试者通常被命名为"participants"（参与者）。有关受试者的信息可能包括研究中使用仪器进行研究和测量的人群、受试者人数、受试者的性别、年龄、文化程度等个人信息，以及选择受试者的原因。

在方法部分，试验的过程经常被描述。读者可以找到关于试验操作步骤的信息。

数据收集也是方法部分的一个重要组成部分。在这一部分，读者将发现与数据收集有关的细节，可能包括数据收集的时间和地点，谁负责收集数据，以及如何收集数据。

在阅读方法部分的数据分析时，读者可以了解分析数据的程序及其可靠性。

二、研究的类型

基于研究目的，可分为基础研究、应用研究和应用基础研究。

基于研究方法，可分为非试验研究、试验研究和准试验研究。非试验研究包括历史研究、描述研究和相关研究。试验研究是根据一些通常被称为"处理变量"的选定标准将参与者分配到组中。准试验研究是根据某些特征或品质，如性别、种族、年龄、邻里等的差异，将参与者预先分配到检测组。研究者对试验对象的分组没有控制权。

以原创性为基础，可分为初级研究和次级研究。初级研究包括直接获取原始数据。初级研究的数据是以前不存在的数据。大多数原始数据是通过试验、问卷调查、个人访谈、邮寄调查和观察收集的。次级研究涉及使用已经存在的数据，可以由现有的文件或其他出版物组成。

三、阅读技巧

在阅读方法部分时，需要确定以下问题：

材料是如何准备的？试验的过程是怎样的？每一步是如何进行的？数据是如何收集和分析的？

这部分通常是用过去时态和被动语态写成的。也有许多现在分词短语和过去分词短语用来表示研究是如何进行的，如介词 by、to 等经常被使用。因此，识别过去时态和被动语态、现在分词和过去分词以及介词短语 by 和 by 中的谓语动词有助于理解这部分。

方法部分经常使用的单词（表 4.4）、短语和句子如下：

（一）常见单词（表4.4）

表4.4　研究型论文研究方法部分经常使用的单词

动词	中文释义	动词	中文释义
analyze	分析；分解；解析	heat	加热
ascertain	确定	highlight	突出；强调；使醒目
assess	评估	identify	识别
calculate	计算	illustrate	说明
clarify	澄清	indicate	表明
consider	考虑	investigate	调查
construct	建造	measure	测量
describe	描述	mix	混合
design	设计	perform	形成
determine	决定	present	呈现
develop	开发	provide	提供
distinguish	区分	stir	搅拌
evaluate	评价	test	测试
examine	检查	understand	理解
experiment	尝试	undertake	承担
explore	探索	use	使用
expose	暴露	validate	验证

（二）常用短语

According to / by / through / adding / allocating / assigning / using / combining / employing / dividing / categorizing / mixing / oscillating / separating...

（三）例句

Gossypium hirsutum 'DP 99B' seeds were disinfected with a mass fraction of 10% H_2O_2

for 30 min, washed three times with distilled water, and then immersed in distilled water for 12 h.

Seeds were germinated in wet sand and then cultivated in a controlled environment chamber under a diurnal cycle of 14h light [450 μmol/ (m^2 · s)] at 30℃ and 10h dark at 25℃.

The data of morphological and physiological indicators were analyzed by ANOVA and multiple comparisons with a significance level of $P<0.05$ using SPSS 22.0 and plotted by Origin 2019b software.

四、举例分析

（一）举例1

Plant growth condition and material preparation. (1) Upland cotton (*Gossypium hirsutum* L.) cv. fuzzless-lintless mutant Xu-142-fl and its wildtype Xu-142, kindly provided by Cotton Research Institute of the Chinese Academy of Agricultural Sciences, were grown at the experimental field of the Henan Institute of Science and Technology, Xinxiang, Henan Province (China). (2) Flowers were tagged on the day of anthesis (0 DPA), -2 and -1 DPA flowers were estimated by the size of flower bud. (3) The bolls of two genotypes were harvested at -2, -1, 0, 1, 2, 5, 10 and 30 DPA and put on ice immediately. (4) Ovules, including -2, -1, 0, 1, 2 DPA of Xu-142 and all investigated fiber development stages of Xu-142-fl, were carefully dissected from each bolls; fiber of Xu-142, including 5, 10 and 30 DPA, was separated from ovule. (5) In addition, 8 days-old cotyledon, the first truly leaf and bud (a diameter of approximately 0.5 to 1 cm) were also harvested. (6) At least three biological replicates were collected for each samples at each developmental stage. (7) All collected samples were frozen in liquid nitrogen immediately and then stored at -80℃ freezer until further use.	句子（1）介绍了试验所使用的试验材料的属性和种植地点。 句子（2）～（5）描述了采样部位和采样时间。 句子（6）说明了重复次数。句子（7）说明了样品保存方法。
RNA isolation and reverse transcription. (8) Total RNAs were extracted from ovules, fibers, cotyledons, leaves and shoot buds using the Biospin Plant Total RNA Extraction Kit (BioFlux) according to the manufacturer's instructions. (9) Briefly, tissues were ground into fine powder in liquid nitrogen, and transfer the pow-	句子（8）～（10）描述了样品中RNA的提取方法。其中，第（11）句中的"by"暗示了RNA提取的完成。

der into 1.5 mL centrifuge tube with lysis and PLANTaid, then immediately vortex several times to mix well and stand at room temperature for 5 minutes. (10) After adding wash buffer, 40 μL DNase Buffer was added to the spin column to purify RNA. (11) The quantity and purity of RNAs were measured by NanoDrop ND-2000 Spectrophotometer (Thermo Scientific, ND-2000, USA).

(12) Based on previous studies, a total of 54miRNAs were selected for this study; these miRNAs are either important to plant development or differential expressed at a certain development stage of cotton differentiation and development. (13) A total of 200 ng total RNAs extracted from difference tissues was used to transcribe into cDNA using TaqMan MicroRNA Reverse Transcription Kit (Applied Biosystems, Foster City, CA, USA) as our previous reports. (14) MiRNA-specific stem loop primer was employed to run reverse transcription. (15) According to the kit supplier's protocol, 15 μL total reverse transcription reaction mixture contains 1.5 μL Reverse Transcription buffer, 1.0 μL MultiScribeReverse Transcriptase, 0.15 μL dNTPs, 0.19 μL RNase inhibitor, and 200 ng total RNAs. (16) When the reaction was done, 80 μL DNase free water was added to dilute the RT-PCR products, and vortex gently to mix well, then store at $-20℃$ for further qRT-PCR.

句子（12）～（16）描述了对样品 RNA 逆转录所采用的方法。其中，第（16）句中明确了 RNA 逆转录的完成。

Quantitative real-time PCR. (17) The real-time RT-PCR was performed with an Thermo Scientific PikoRea Real-Time PCR System (Thermo Scientific, PikoReal Real 96, USA). (18) According to our reported previously, each reaction includes 1 μL cDNA, 2 μL forward and reverse primers, 5 μL 2× SYBR green mixture and 2 μL DNase free water. (19) The qRT-PCR was performed as following: 10 min at 95℃, followed by 40 cycles of 15 s at 95℃ and 1 min at 60℃. (20) Based on our previous study, UBQ14 is the most reliable reference gene in cotton; thus we used it as reference gene to normalize the tested gene expression value. (21) SPSS was employed to analyze the biostatistics significance. (22) A heat map was generated with MeV (Multi Experiment Viewer) according to previous reported (Sun et al. 2015a, b).

句子（17）～（20）描述了对 cDNA 定量所采用的方法。

句子（21）～（22）说明了统计和作图的方法。

Predication of miRNA targets and function analysis. (23) The mature miRNA sequences and the CDS sequences that extracted from CottonGen database (http://www.cottongen.org) were submitted to psRNATarget database (http://plantgrn.noble.org/psRNATarget/) to search

句子（23）～（24）说明了 miRNA 靶标的预测方法。

miRNA targets. (24) The optimized criteria as follows: the Hspice value in the range from 1 to 17 nt; the central mismatch between 10 and 11 nt; the rest parameters were defined by the program.

(25) Identified miRNA targets were performed alignment against KEGG Orthology (KO) and EuKaryotic Orthologous Groups (KOG) database for Kyoto Encyclopedia of Genes and Genomes (KEGG) pathway enrichment and KOG term classification. (26) To visualize the biological process, cellular component and molecular function that miRNA targets involved in, the NR database and Blast2go software were employed to analyze miRNA targets Gene Ontology (GO) term classification.

句子（25）~（26）说明了miRNA靶标功能的分析方法。

（二）举例2

Cotton cultivars. (1) Four commercialized transgenic insect-resistant cotton cultivars, CCRI41, DP 99B, CCRC21 and BAI1, were used in this study. (2) CCRI41 was bred by the Cotton Research Institute of the Chinese Academy of Agricultural Sciences; CCRC21 was bred by the Cotton Research Center of the Shandong Academy of Agricultural Sciences; DP 99B was bred by Monsanto Company; and BAI1 was bred by Henan Institute of Sciences and Technology.

句子（1）~（2）介绍了试验所使用的试验材料及其选育单位。

(3) In China, transgenic insect-resistant cotton cultivars are widely adopted in the yield in recent years. (4) Maybe, insect-resistant gene transformation could result in the different responses of cotton plants to mechanical stress, but it was not in the research scopes of this experiment. (5) Therefore, only transgenic cotton cultivars were used in this experiment. (6) All four cultivars have growth periods of around 130 days. (7) However, CCRI41 and DP 99B cultivars often pre-maturely senesced, BAI1 was resistant to pre-mature senescence, and CCRC21 was between them.

句子（3）~（7）说明了采用转基因抗虫棉材料的原因和特点。

Field experiment. (8) The four cotton cultivars were planted in a sandy loam soil with a pH of 8.5 (water : soil = 5 : 1), an organic matter content of 0.60% (determined by digestion with potassium dichromate under strongly acidic conditions), an available nitrogen content of 18.6 mg/kg (determined by extraction with 1 M KCl), an available P content of 16.2 mg/kg (determined by extraction with 0.5 M $NaHCO_3$), and an available K content of 158.5 mg/kg (determined by extraction with 1 M NH_4OAC).

句子（8）描述了试验材料的培养方法。

(9) Planting was conducted in 2009 and 2010 at the experimental field station (35°16′N; 113°56′E) of the Henan Institute of Science and Technology, Xinxiang, Henan Province.

(10) The cotton seeds were sowed under plastic film mulching, respectively, on April 26th, 2009 and 2010, according to a random block design with four biological replicates. (11) Each block contained 4 plots, and each plot was planted with only one cultivar. (12) Each plot contained four 10 m long rows with an inter-row spacing of 0.8 m and an intra-row spacing of 0.27 m. (13) The planting density was 45,000 plants per hectare.

(14) Conventional agricultural practices were applied in the study. 150 kg N, 100 kg P and 75 kg K in the form of urea, diammonium phosphate and potassium sulphate, respectively, were applied per hectare before sowing. In the early flowering stage, additional quantities of urea-N (150 kg N per hectare) and K (75 kg K per hectare) were applied by top-dressing. (15) All plots were treated with chemical pesticides to keep insects away.

Hanging labels from flower petioles (HLFP). (16) During early flowering season, 5 adjacent plants in one of the central rows of each plot were selected and tagged with labels that were hung from their flower petioles during 9:10 A.M every day and kept in place throughout the blossoming stage. (17) HLFP was made for each flower at its opening day. (18) The anthesis and boll-opening dates for the tagged plants were recorded on their labels.

(19) A label with a thin thread at its one end weighted 337±20 (mean±SD) mg with length of 3.5 cm, width of 3.0 cm and thickness of 2.2 mm. (20) The label was tied to flower petiole by thin thread and vertically suspended when tagging. (21) The labels are widely used in field experiment of cotton.

Measurement of plant shape parameters and yield and fiber quality traits. (22) During the harvest season, the numbers of fruiting branches (FB) and fruiting positions (FP) on each plant were counted; the height of the stem (from base to tip) was measured, along with the length of each fruiting branch and the location of each fruiting position. (23) The fruiting branches were recorded as the 1st FB, 2nd FB, *etc*, from the bottom. (24) The fruiting positions were numbered horizontally, as the 1st FP, 2nd FP, etc.

from the stem. (25) The fruiting position length was defined as the distance between the 1st FP and the main stem or between adjacent FPs; the distance between the stem and the 1st FP was recorded as the 1st FP length (FPL), while that between the 1st and 2nd FPs was recorded as the 2nd FPL, and so on. (26) The fruiting branch length (FBL) was calculated as the sum of all FPLs on a single branch, $i.e.$, the distance between the main stem and the terminal FP.

(27) We defined four "layers" of fruiting branches: the first layer consisted of the 1st to 4th FBs, the second layer consisted of the 5th to 8th, the third layer consisted of the 9th to 12th, and all FBs from the 13th upwards were assigned to the fourth layer. (28) All fruiting positions more distal than the 3rd FP on a given fruiting branch were collectively referred to as >3rd FPs. (29) The number of bolls, fruiting branches and fruiting position lengths within each FB layer were calculated. (30) Since many plants had fewer than 13 fruiting branches and many fruiting branches had fewer than three fruiting positions, data for the fourth fruiting branch layer and for fruiting position numbers greater than three on any given branch were not subjected to statistical analysis.

(31) Cotton was harvested twice, on September 22nd and October 22nd. (32) All opened bolls were harvested and counted. (33) The harvested cotton was weighed as seed cotton yield per plant. (34) Boll weight was average seed cotton weight of all harvested opened bolls in one plant. (35) Ginned cotton was used for determination of fiber quality traits such as fiber length, fiber strength, micronaire, uniformity index and fiber elongation with high volume instrument (HVI).

句子(31)~(35)说明了测定的籽棉和皮棉产量所使用的方法。

Biostatistical analyses. (36) The experiment was first conducted in 2009 and then repeated in 2010, with both replicates yielding similar results. (37) In both cases, 20 control and treated plants of each genotype were studied, and were distributed evenly among the four experimental plots. (38) The two-year experimental results were pooled and subjected to statistical analysis by means of analysis of variance (ANOVA), and Student's t-test was used to identify significant differences (at the $P < 0.05$ level) between the mean results for different treatments and those observed for control plants.

句子(36)~(38)说明了所使用的数据分析方法。

阅读练习

阅读以下研究型论文的试验方法，并完成表 4.5 列出的任务。

Plant materials

（1）In 2008, extreme low-temperature and rainy disaster weather occurred over a ten-day period in the middle of January and another ten-day period in the middle of February in Nanning, Guangxi Province. （2）In March, the overwintering survival rate of the annual cultivar H077（*G. hirsutum* L.）and perennial germplasm H113（*G. barbadence* L.）planted in the open field was 0 and 100%, respectively（Zhang et al., 2008）.

（3）Here, H077（weak cryotolerance）and H113（strong cryotolerance）were used as P_1 or P_2. （4）Six populations of different generations, i.e., P_1, P_2, F_1, B_1, B_2 and F_2, from each of the two reciprocal crosses H077 × H113（♀ × ♂, similarly hereinafter）and H113 × H077 were composed for genetic analysis. （5）In cross H077 × H113, P_1, P_2, F_1, B_1, B_2 and F_2 represented H077, H113, H077 × H113, （H077 × H113）× H077, （H077 × H113）× H113 and （H077 × H113）×（H077 × H113）, respectively; in cross H113 × H077, P_1, P_2, F_1, B_1, B_2 and F_2 represented H113, H077, H113 × H077, （H113 × H077）× H113, （H113 × H077）× H077 and （H113 × H077）×（H113 × H077）, respectively.

Field experiment design

（6）The field trial was carried out in the experimental field of Guangxi University（22°56′N, 108°21′E）, where the annual mean temperature was 21.6℃, the average temperature of the hottest month（July）was 28.3℃, the average temperature of the coldest month（January）was 12.3℃; ≥10℃ active accumulated temperature amounted to 7,370.5℃, and the average precipitation was 1,300.6 mm. （7）The soil of experimental plot was krasnozem.

（8）Seeds of plant materials mentioned in above were sowed on March 24th, and the testing seedlings were transplanted to the field on April 8th, 2009. （9）For each cross, generations of P_1, P_2 and F_1 were planted in 2 rows, generations of B_1 and B_2 were planted in 10 rows, and generation of F_2 was planted in 12 rows. （10）The field trial was a random block design with 3 replications. 5 plants per row; length and width of each row were 4.0 m and 0.65 m, respectively.

Determination methods

(11) Relative conductivity was always chosen as the first physiological parameter when cryotolerance was investigated (Deans et al., 1995; Eugénia et al., 2003; Murray et al., 1989; Paul et al., 2006), since that it is quick, easy, user independent, cheap and quantitative for identifying the cryotolerance of large populations (Guo et al., 2009). (12) Therefore, it was measured as the index of cryotolerance in this research.

(13) The similar branches of each plant were taken into frige and treated with 0℃ on January 28th, 2010. (14) After 24 hours, the barks of the branches were peeled and cut into chippings, then their relative conductivity were measured by using an electric conductance meter as described by Dionisio-Sese and Tobita (1998).

Data analysis

(15) Suppose under the modifications of polygene and environment, the effect of major gene in segregation generation displayed an independent normal distribution, and then the whole segregation generation could be seen as a mixed distribution made up by many independent normal distributions. (16) The joint analyses of P_1, P_2, F_1, B_1, B_2 and F_2 populations might involve the following genetic models: 1 major gene (A model), 2 major genes (B model), polygene (C model), 1 major gene and polygene (D model), 2 major genes and polygene (E model). (17) Stastical deduction could be regarded as an estimation of probability distribution of observed value according to the principle of entropy maximization suggested by Akaike (Gai et al., 2003). (18) Therefore, comparison of the goodness-of-fit of practical frequency distributions with the standard curves can approximate the genetic model of a quantitative trait. (19) The most-fitting genetic model was chosen according to Akaike's Information Criterion (AIC). (20) The smallest AIC value was the most-fitting genetic model. Akaike (Gai et al., 2003) suggested that the hypothesis maximizing the expected entropy should be selected as the best fitting model. For this purpose, based on goodness of fit and parsimony, the hypothesis leading to the smallest AIC would be chosen. (21) The analysis software was provided by Dr. ZHANG Yuan-Ming, and the genetic parameters were estimated with the method, referred to Gai et al., (2003).

(22) For a mixed genetic model, the phenotypic value (p) could be expressed as the summation of population mean (m), major gene effect (g), polygene effect (c) and environmental effect (e), i.e., $p = m + g + c + e$, where g is varied with different major gene geno-

types, and *c* and *e* were normally distributed variables. (23) So, the phenotypic variation (σ_p^2) could be expressed as major gene variation (σ_{mg}^2), polygene variation (σ_{pg}^2) and environmental variation (σ_e^2). $\sigma_p^2 = \sigma_{mg}^2 + \sigma_{pg}^2 + \sigma_e^2$. (24) Therefore, we could define the heritability of major gene (h_{mg}^2) and the heritability of polygene (h_{pg}^2) as h_{mg}^2 (%) = $\sigma_{mg}^2/\sigma_p^2 \times 100\%$ and h_{pg}^2 (%) = $\sigma_{pg}^2/\sigma_p^2 \times 100\%$, respectively.

表 4.5　研究型论文试验方法阅读练习

不同作用的句子	序号
（1）描述所用材料的句子	
（2）显示试验过程的句子	
（3）反映了所使用工具的句子	
（4）说明如何进行测量的句子	

第五节　研究结果

一、概述

研究型论文的结果部分是以图形和表格的形式呈现研究要点。最好是按照时间顺序和有序的顺序，让读者了解研究的结果，而不必演绎、解释或讨论研究结果。因此，研究结果部分通常是一篇研究型论文中最短的部分，但却是研究型论文中最有价值的部分，研究结果揭示了论文的学术价值。

二、特点

结果部分通常具有以下特点：

（一）客观

本部分没有讨论或演绎或解释。

（二）逻辑性强

遵循研究主题的顺序。

（三）简洁明了

在本部分中没有重复的数据和相同的信息。

（四）使用非言语的语言

经常使用图表呈现研究结果。

（五）使用混合时态

通常，假设检验研究的结果和新方法的测试报告用过去时态；描述研究的结果通常用现在时态。解释图表的陈述和总结方法的陈述通常用现在时态，而呈现最重要的发现的陈述通常用过去时态。

三、语步

在阅读研究型论文的结果部分时，您可以关注3个语步。

（一）提供预备信息

在这一举措中，作者为结果的呈现提供了相关的信息，从而将不同的部分连接起来。

（二）用图表描述数据

在这一举动中，作者描述了从试验中得到的图表数据。

（三）基于数据报告结果

在这一步中，作者以一种没有偏见或解释的逻辑顺序呈现研究结果，为读者在讨论部分的解释和评价做好准备。

四、图表

"好文配好图,好图抵千言",科研论文中插图的重要性不言而喻。在研究型论文中,经常使用图形和表格以可视化的方式表示数据,以提高读者对结果及其意义的理解。

图和表按顺序单独并连续编号,从图 1 和表 1 开始。可以在 Results 部分的末尾找到所有的说明性材料,尽管在大多数情况下它们都嵌入到文本中,避免将文本分割成小块。

图和表的标题都应与图或表的宽度相匹配。图的标题位于图下方,而表的标题位于表上方。下面举例说明一些论文中常用的图表。

(一)插图分类

SCI 论文插图可分为三大类,分别是照片(通过照相机、扫描仪或 CCD 成像设备获得)、线条图,以及表达文字意思的示意图。

1. 照片

照片类插图对分辨率的最低要求是:彩图大于 300 dpi;黑白/灰度图大于 600 dpi。为了应对不同期刊对插图处理的各种需求,我们在采集原始照片的时候应尽可能使用高分辨率模式成像(建议长宽比不低于 2 560 × 1 920 dpi,一部 500 万像素以上配置的相机即可满足要求)(图 4.2)。

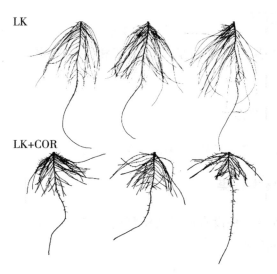

图 4.2　SCI 论文插图(照片示例 1)

所有光学显微镜照片（细胞、荧光）、电子显微镜照片、电泳凝胶照片等，以及其组合图都属于照片类插图（又叫位图），一般要求用 TIFF 格式（或高分辨率 JPG 格式）保存，投稿时多与论文正文分开上传（图 4.3）。

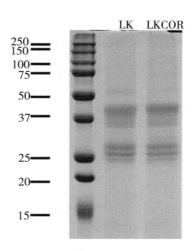

图 4.3　SCI 论文插图（照片示例 2）

照片在作为论文插图之前需要预处理，如瑕疵修复、调整光线、对比度，长宽比/大小（插图尺寸要求，后面介绍）等，通常使用 Adobe Photoshop 软件来完成，其处理过程中可保持图片分辨率且不丢失像素信息。

2. 线条图

线条图包括各类采样软件输出的结果，如红外热成像图、流式细胞图、色谱图、质谱图等，以及通过原始数据统计分析后制作的条形图、折线图、饼图、散点图、箱线图、小提琴图、热图、韦恩图、雷达图、玫瑰图、聚类图等。这类图建议保存为 PDF 格式（矢量图特性：通过数学公式记录点、线及颜色信息，无论放大多少倍都非常清晰，色彩也不会失真）。PDF 格式的图片可通过 Adobe Illustrator 软件随意编辑和优化。需要注意的是，无论哪种方式获得的线条图，最好不要通过截图软件来保存原始结果，这样做将极大损失图像的像素信息，而且可能会导致后期图片无法修改。

（1）色谱图和质谱图

LC-MS/MS 作为代谢组学和蛋白组学分析的主要手段，所分析的样品分子过于微小肉眼不可见，需要借助色谱图、质谱图判断其表现。色谱图评价的是母离子在色谱上的表现，质谱图则是一级母离子和二级碎片子离子在质谱里的信号表现。区分这两种图谱最简单的方法是看横坐标，横坐标是时间轴的是色谱图，横坐标是质荷比的那就是质谱图了，不管色谱图还是质谱图，纵坐标都是信号强度。

常见的色谱图有 Basepeak 图、TIC 图、XIC 图；质谱图经常用的是一级质谱图，二级质谱图，b，y 离子匹配图（有注释信息的二级质谱图），下面是某代谢物样品 HILIC-正离子模式 TIC 图谱（图 4.4）。

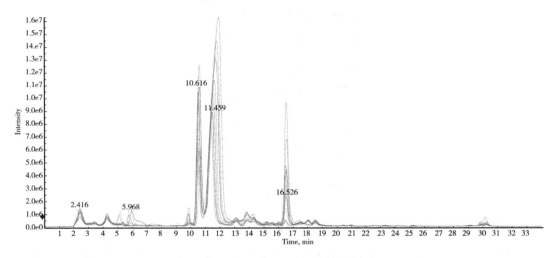

图 4.4　SCI 论文插图（色谱图示例）

（2）条形图

条形图是由长度与其所代表的值成比例的矩形组成的图（图 4.5）。条形图可以垂直或水平绘制。例如，图 4.5 展示了低钾（Low K，LK）条件下冠菌素（COR）对棉花幼苗叶面积、根系表面积、茎长、根长、根粗和根体积生长的影响。深黑色条和灰色条分别代表低钾 LK 和 LK+ COR 处理。图 A 是普通的并列条形图；图 B 在纵轴上有一个折断（为了便于同时展示数据差别较大的两个性状）；图 C 有两个纵轴（两个性状的量纲不一致）。

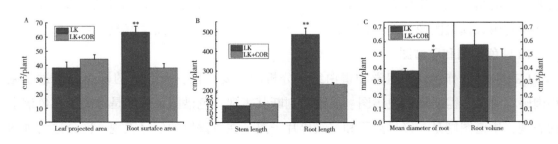

图 4.5　SCI 论文插图（条形图示例）

（3）饼图

饼图是一个圆形的统计图（图4.6），分成几个部分来说明数字的比例。在饼图中，每个切片的弧长与它所表示的数量成正比。通常用于表示百分比。例如，在下面的饼图中，每一片扇叶代表一类功能相似的蛋白质百分比，所有的扇叶构成了在冷胁迫和嫁接诱导下差异表达的68种蛋白质。

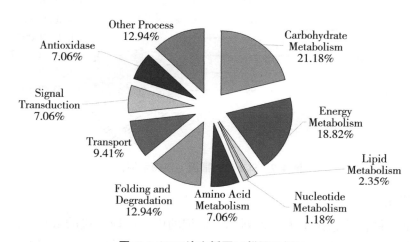

图 4.6　SCI 论文插图（饼图示例）

（4）热图

热图是一种很常见的图，其基本原则是用颜色代表数字，让数据呈现更直观、对比更明显。常用来表示不同样品组代表性基因的表达差异、不同样品组代表性化合物的含量差异、不同样品之间的两两相似性。实际上，任何一个表格数据都可以转换为热图展示。

热图通过将数据矩阵中的各个值按一定规律映射为颜色展示，利用颜色变化来可视化比较数据。当应用于数值矩阵时，热图中每个单元格的颜色展示的是行变量和列变量交叉处的数据值的大小；若行为基因，列为样品，则是对应基因在对应样品的表达值；若行和列都为样品，展示的可能是对应的两个样品之间的相关性。

数字映射到颜色可以分为线性映射和区间映射。线性映射是每个值都对应一个颜色，区间映射是把数值划分为不同的区间块，每个区间块的所有数字采用同一个颜色显示。两者没有优劣好坏之分，具体使用取决于展示意图。下图展示了不同处理下棉花幼苗木质部汁液中的13种双功能抑制剂/脂质转移蛋白/种子贮藏2S白蛋白超家族蛋白的表达量，并分析了它们的进化关系（图4.7）。

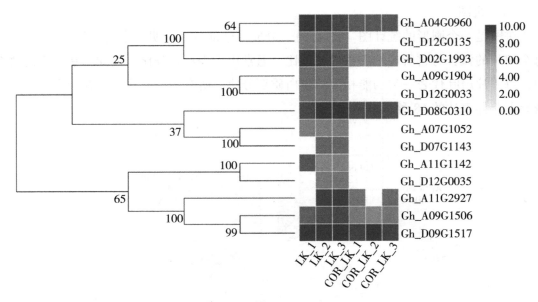

图 4.7　SCI 论文插图（热图示例）

（5）折线图

折线图是一种将信息显示为一系列称为标记的由直线段连接的数据点图。折线图是许多领域中常见的一种基本的图，特征是显示了一种通常发生在一段时间内的模式或趋势。在下面的示例（图 4.8）中，图中的直线表示利用 ΔK 检测陆地棉（*Gossypium hir-*

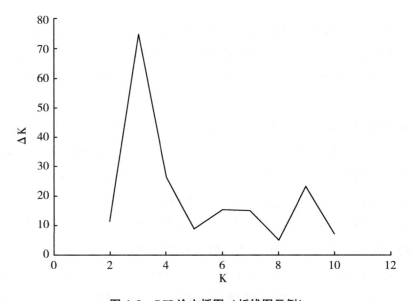

图 4.8　SCI 论文插图（折线图示例）

sutum）种群中的亚群（K）数量。K（ΔK）是一个特设量，基于数据的对数概率相对于子种群数量的二阶变化，K 的范围从 1 到 10。可以看到 K=3 时，ΔK 出现峰值，说明所研究的陆地棉群体可以分成 3 个亚群。

（6）散点图（火山图、曼哈顿图）

散点图通常包含 5 个点以上的数据。散点图类似于线图，使用水平轴和垂直轴来展示数据点。然而它有一个非常具体的目的，散点图显示一个变量受另一个变量影响的程度，变量之间的关系称为它们之间的相关性。当数据点与回归直线越接近时，说明变量之间的相关性越高，或者关系越强。如果一个变量值随另一个变量值的增大而增大，那么这两个变量就被认为是正相关的。如果一个变量值随另一个变量值的增大而减小，则这两个变量具有负相关性（图 4.9）。

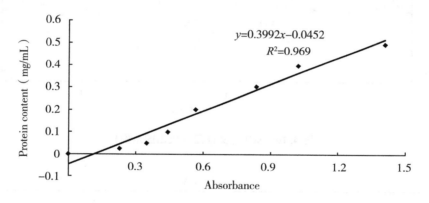

图 4.9　SCI 论文插图（散点图示例）

火山图是散点图的一种，将统计分析中的显著性量度（如 P 值）和变化幅度相结合，从而能够帮助研究者快速直观地识别那些变化幅度较大且具有统计学意义的数据点（如代谢物等）。火山图是一种单变量统计分析方法，常应用于研究基因组、转录组、代谢组、蛋白质组等数据分析。图 4.10 是某样品 HILIC 正离子模式下数据结果的火山图，颜色较深的点为显著性差异代谢物（FC > 2.0，P < 0.05）（图 4.10）。

曼哈顿图，因其形似曼哈顿摩天大楼，故俗称为曼哈顿图。曼哈顿图本质上是散点图，一般用于全局展示大量非零数据的波动，y 轴高点显示出具有强相关性的位点，最早应用于全基因组关联分析（Genome-wide association study, GWAS）研究中，可以一目了然地看基因的频率、分布，同时可以知道目标位点的具体位置及显著性（图 4.11）。

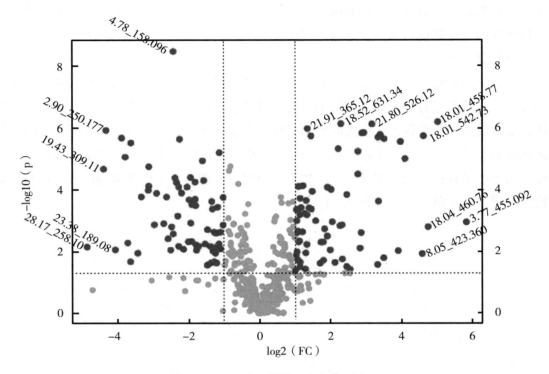

图 4.10　SCI 论文插图（火山图示例）

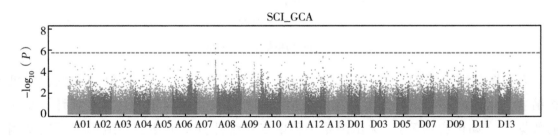

图 4.11　SCI 论文插图（曼哈顿图示例）

（7）箱线图和小提琴图

一般来说，小提琴图是一种绘制连续型数据的方法，可以认为是箱线图与核密度图的结合体。在小提琴图中，我们可以获取与箱线图中相同的信息：

①中位数（小提琴图上的一个白点）。

②四分位数范围（小提琴中心的黑色条）。

③较低/较高的相邻值（黑色条形图）——分别定义为下四分位数和上四分位数。

这些值可用于简单的离群值检测技术，即位于这些"栅栏"之外的值可被视为离群值。与箱线图相比，小提琴图毫无疑问的优势在于：除了显示上述的统计数据，还显示了数据的整体分布。这个差异点很有意义，特别是在处理多模态数据时，即有多峰值的分布（图4.12）。

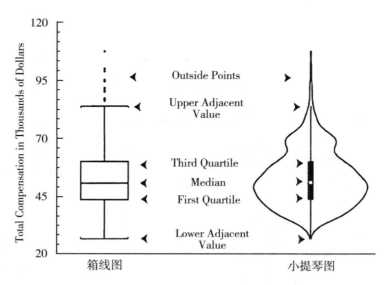

图 4.12　SCI 论文插图（箱线图和小提琴图示例）

（8）气泡图（点阵图）

上文的散点图展示了两个数值变量之间的关系。气泡图将数据点替换为气泡，用气泡大小表示第三个变量维度。气泡图是以一系列气泡的形式直观地表示信息的图。这种类型的图可以用来比较概念，识别相似和不同的领域，并提出各种各样的信息活动的目的，如制作演示文稿，规划设计，并制定战略。下面的气泡图显示了 26 个匹配的京都基因和基因组百科全书（KEGG）通路的通路影响图。颜色越深表示 P 值越低，气泡越大表示影响力越高（图4.13）。

（9）雷达图

雷达图也称为蜘蛛图、星图、蜘蛛网图，被认为是一种表现多维数据的图。雷达图将多个维度的数据量映射到坐标轴上，每一个维度的数据都分别对应一个坐标轴，这些坐标轴以相同的间距沿着径向排列，并且刻度相同。连接各个坐标轴的网格线通常只作为辅助元素，将各个坐标轴上的数据点用线连接起来就形成了一个多边形。坐标轴、点、线、多边形共同组成了雷达图（图4.14）。

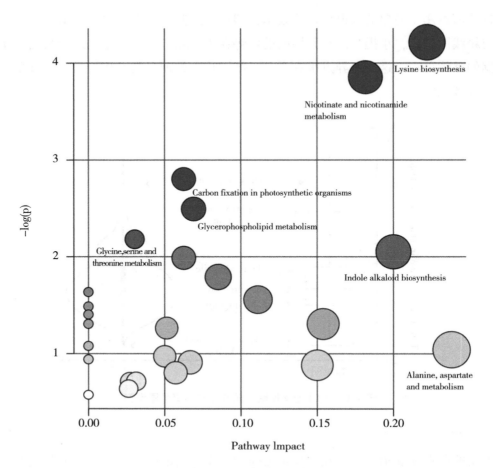

图 4.13　SCI 论文插图 [气泡图（点阵图）示例]

3. 示意图

示意图的目的是更直观地表达研究思路、说明筛选流程、综合研究结果等，使之简明易懂。如试验过程示意图、信号转导通路图、研究流程图、机制模式图等。这类图片最好通过 Adobe Illustrator 软件绘制，有的作者也通过 Office（Excel、PPT 或 Word）、Visio、亿图等软件绘制示意图，但不论采取哪种途径，结果最好都保存为 PDF 或 EPS 格式（即矢量图格式），以便后期通过 Adobe Illustrator 修改和按照不同期刊的投稿要求制作插图。下面以流程图示例。

流程图是一种表示算法、工作流程或过程的图表类型，流程图将工作流程显示为各种类型的图或框，并通过箭头将它们连接起来显示它们的顺序。流程图用于分析、设计、记录或管理各个领域的过程或程序。下面是美国应用生物系统公司基于八通道 iTRAQ 标记检测差异蛋白质的一个试验流程（图 4.15）。

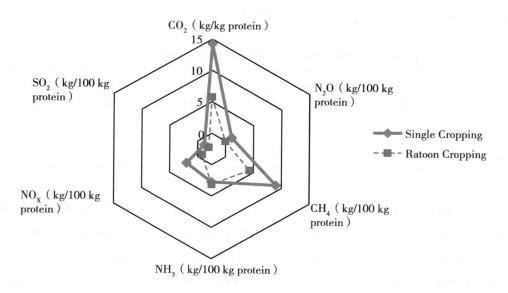

图 4.14　SCI 论文插图（雷达图示例）

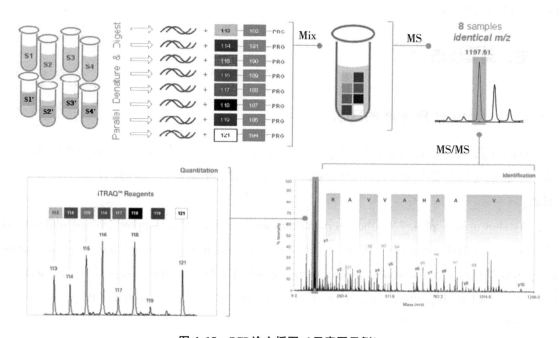

图 4.15　SCI 论文插图（示意图示例）

（二）表格

表格是一组按列和行排列的书面事实和数字，是研究型论文中组织数字信息的一种非常有用且常用的方法。下面表4.6是一个例子。

表4.6 Qualification and quantification of xylem sap proteins.

Identified proteins	Qualitative proteins	Quantitative proteins compared	Non-quantitative proteins compared		
			ULK	ULK+COR	Others
Total	516	356	44	22	94
Uncharacterized	8	0	2	0	6

ULK, undetectable in the low K (potassium) treatment; ULK+COR, undetectable in the low K with COR (coronatine). Raised cotton seedlings in wet sand were transferred into a normal solution, grown for 3 days, then separated into a K-deficient solution or a new normal solution, and both treatment groups were grown for 4 days. These cotton seedlings were used to sample the xylem sap; three biological replicates of samples were used for protein identification, qualification, and quantification.

五、阅读技巧

(一) Data 与 Result 的区别

Data（数据）是直接从试验或观察中获得的事实、数字或细节，没有一般的陈述或描述。数据可以是原始数据、汇总数据（平均值和标准偏差）或转换数据（百分比）。例如：

As shown in（表4.7）, the overwintering survival rate of all grafted groups under field condition were higher (mean, 10.44%) than that of the control (self-grafted A4), and grafted plants "A4/F113" was best among them (reached 100%); the height of the highest regeneration bud of all grafted groups were higher (mean, 15.75%) than that of the control, except grafted plants "A4/P098".

表4.7 Analysis on overwintering survivorship of grafted plants in the open field.

Scion/ rootstock	Overwintering survival rate			Height of regeneration bud		
	ADV (%)	SFV	ADV (cm)	SFV	$\overline{U}_{os}$	Order
A4/F118	96.67±3.33	0.75	61.72±2.71	0.91	0.83	2

(续表)

Scion/rootstock	Overwintering survival rate			Height of regeneration bud		
	ADV (%)	SFV	ADV (cm)	SFV	$\overline{U}_{os}$	Order
A4/F697	96.67±3.33	0.75	59.86±6.06	0.77	0.76	4
A4/F098	96.67±3.33	0.75	60.82±8.36	0.84	0.80	3
A4/F112	93.33±3.33	0.50	55.13±2.59	0.41	0.46	5
A4/F113	100.00±0.00	1.00	62.88±8.71	1.00	1.00	1
A4/P098	96.67±3.33	0.75	49.73±2.06	0.00	0.38	6
A4/P113	90.00±5.78	0.25	55.50±4.30	0.44	0.34	7
A4/A4 (CK)	86.67±8.82	0.00	50.07±5.62	0.03	0.01	8

ADV means actual determination value; SFV means subordinate function value; $\overline{U}_{os}$ is the abbreviation of the average SFV of overwintering survivorship.

doi: 10.1371/journal.pone.0063534.t003

Result（结果）是描述数据的一般性陈述，实际上是摘要、转换数据，有时还以文本的形式表达出来。因此，理解插图和表格中的信息是很重要的。例如：

When exposed to 0℃ for 48 h, the relative conductivity, content of soluble protein and free proline in all shoot barks were heightened greatly, but the soluble sugar content was greatly reduced, when compared with exposure to 20℃ for 48 h.

通常，数据和结果在呈现结果时混合在一起。例如：

After chilling treatment, it was shown that the relative conductivity of the grafted plants in shoot bark was lower (8.80%), and the content of soluble sugar, soluble protein and free proline which were higher (25.00%, 1.55%, 14.51%, 3.46%, respectively) than that of the control (表4.8).

表4.8 The physiological parameters change rate of grafted A4 after chilling treatment

Parameter	Relative conductivity (%)	Soluble sugar (%)	Soluble protein (mg/g)	Free praline (mg/g)
mean of not-self grafted A4	53.08±4.15	0.40±0.04	1.31±0.20	20.63±0.72
Self-grafted A4 (CK)	58.20±2.91	0.32±0.13	1.29±0.10	19.94±0.53
Change rate by not-self grafting	-8.80%	25.00%	1.55%	3.46%

doi: 10.1371/journal.pone.0063534.t001

(二) 惯用语

有一些惯用语可以用来帮助您识别数据的描述和结果的陈述。

1. 用于描述数据的惯用语

(1) ...form/comprise/make up/constitute/account for...percent.

(2) A is...times as much/many as B.

(3) The pie chart consists of...segments, the largest one representing...which accounts for...of the total.

(4) The number of...increased by...

(5) ...add up to...

(6) Compared with...increases/decreases about...

(7) ...experience an increase/a decline of...

(8) ...remain steady/level/unchanged.

2. 用于结果陈述的惯用语

(1) In our experiments, we found...

(2) We found that...First...Second...

(3) One of the key findings of this study was...

(4) The answer has been demonstrated in two ways.

(5) It was found/observed/noted/seen/recorded that...

(6) The results/This section showed/described/indicated/revealed/detected.

3. 图表描述词汇、词组及句型常用表达

基本要素类型：table, chart, diagram/graph, column chart, pie graph

描述：show, describe, illustrate, reveal, represent, can be seen from, clear, apparent

内容：figure, statistic, number, percentage, proportion

表示数据变化的单词或者词组动词：

(1) 起伏、波动

rise and fall, fluctuate, wave, undulate, rebound, recover

(2) 高、低点值

peak, reach a (high) peak/point, reach the bottom, reach a low point

(3) 增加、上升、提高

increase, grow, rise, climb, expand, ascend, skyrocket, soar

（4）减少、下降

decrease, decline, fall, drop, descend, diminish, slide, shrink, collapse

4. 形容词或副词

（1）变化巨大

迅速的，飞快的，rapid/rapidly

巨大的，引人注目的，dramatic/dramatically

有意义的，重大的，significant/significantly

锐利的，明显的，sharp/sharply

急剧升降的，steep/steeply

（2）变化平缓

稳固的，平稳的，steady/steadily

渐进的，逐渐的，gradual/gradually

缓慢的，不活跃的，slow/slowly

轻微的、略微的，slight/slightly

稳定的，stable/stably

（三）英语图表写作套句推荐

1. 图表的总体描述

（1）The table shows the changes in the number of... over the period from... to...

（2）The data/statistics/figures lead us to the conclusion that...

（3）As can be seen from the diagram, great changes have taken place in.../ the two curves show the fluctuation of...

（4）From the table/chart/diagram/figure, we can see clearly that... or it is clear/apparent from the chart that...

（5）This is a graph which illustrates...

（6）The graph... presented in a pie chart, shows the general trend in...

（7）This is a column chart showing...

2. 时间段表达方式

（1）over the period from... to... the... remained level.

（2）in the year between... and...

（3）in the 3 years spanning from 1995 through 1998...

（4）from then on/from this time onwards...

3. 数据变化表达方式——表示上升趋势的句型

（1）the number sharply went up to...

（2）the figures peaked at... in (month/year)

（3）the situation reached a peak (a high point at) of...

（4）a increased by...

（5）a increased to...

（6）there is an upward trend in the number of...

（7）a considerable increase/decrease occurred from... to...

（8）...(year) witnessed/saw a sharp rise in...

4. 数据变化表达方式——表示下降趋势的句型

（1）the figures/situation bottomed out in...

（2）the figures reached the bottom/a low point/hit a trough.

（3）from... to... the rate of decrease slow down.

（4）from this year on, there was a gradual decline/reduction in the... reaching a figure of...

5. 数据变化表达方式——表示平稳趋势的句型

（1）the number of... remained steady/stable from (month/year) to (month/year).

（2）the percentage of... stayed the same between... and...

（3）the percentage remained steady at...

6. 表示比较的句型

（1）the percentage of... is slightly larger/smaller than that of...

（2）there is not a great deal of difference between... and...

（3）... decreased year by year while... increased steadily.

（4）there are a lot similarities/differences between... and...

（5）A has something in common with B.

（6）the difference between a and b lies in...

7. 表示倍数的句型

（1）the graphs show a threefold increase in the number of...

（2）A is... times as much/many as B.

六、举例分析

（一）Effect of Hang labels from flower petiole (HLFP) on cotton plant shape

（1）As shown in Table 4.9, HLFP significantly reduced the stem height in all four genotypes relative to control plants by between 7.8% and 36.5%. （2）HLFP also significantly reduced the fruiting branches (FB) number for the DP 99B variety, but had no significant effect on FB numbers in other genotypes. （3）Furthermore, HLFP significantly reduced the total number of fruiting positions in the CCRI 41, DP 99B and CCRC 21 varieties by 33.3%, 37.4% and 18.8%, respectively, but had no effect on the fruiting position count in BAI 1 (表 4.9).

句子（1）~（3）描述了由表 4.9 数据所揭示的结果或发现。

表 4.9 Plant height, fruiting branches, and fruiting positions of four cotton genotypes under mechanical stimuli- "hang labels from flower petiole (HLFP)" for every white flower in the field. For each genotype, means at the same column with different lower cases denoted significant difference between CK and HLFP at $P<5\%$, $n=20$ plants (The same below).

Genotypes	Treats	Plant height (cm)	Fruiting branches (No./plant)	Fruiting positions (No./plant)
CCRI41	Control	96.6 a	14.4 a	59.9 a
	HLFP	68.5 b	13.1 a	40.2 b
DP 99B	Control	86.4 a	15.1 a	78.0 a
	Control	86.4 a	15.1 a	78.0 a
BAI1	Control	94.7 a	13.1 a	52.6 a
	HLFP	79.3 b	13.1 a	47.9 a
CCRC21	Control	96.8 a	13.0 a	51.5 a
	HLFP	89.2 b	12.4 a	41.8 b

https://doi.org/10.1371/journal.pone.0082256.t001

（二） Effects of HLFP on cotton plant yield and fiber quality

（4） In the CCRI 41 and DP 99B varieties, HLFP significantly reduced the number of bolls per plant by 34.8% and 27.6%, respectively. （5） However, it did not affect boll numbers in BAI 1, and significantly increased them for CCRC 21. （6） HLFP significantly reduced boll weights by between 10.6% and 14.3% in all four varieties, and also significantly reduced the seed cotton yields for the CCRI 41, DP 99B and BAI 1 varieties by 44.3%, 38.5% and 9.3%, respectively. （7） However, it significantly increased the cotton seed yield by 11.2% for CCRC 21 （表4.10）.

> 句子（4）~（7）描述了由表4.10数据所揭示的结果或发现。

表 4.10 Effects of mechanical stimuli- "hang lables from flower petiole (HLFP)" for every white flower on the number of bolls set, boll weight and seed cotton yield of four cotton genotypes in the field.

Genotypes	Treats	Bolls (No./Plant)	Boll weight (g)	Seed cotton yield (g/plant)
CCRI41	Control	20.4 a	4.9 a	99.5 a
	HLFP	13.3 b	4.2 b	55.4 b
DP 99B	Control	23.9 a	4.9 a	117.0 a
	HLFP	17.3 b	4.2 b	72.0 b
BAI1	Control	16.6 a	4.4 a	73.8 a
	HLFP	17.0 a	3.9 b	66.9 b
CCRC21	Control	13.0 b	4.7 a	61.6 b
	HLFP	16.3 a	4.2 b	68.5 a

https：//doi.org/10.1371/journal.pone.0082256.t004

 阅读练习

1. 阅读以下研究型论文的结果，并尝试找到结果的描述词

Distributions of relative conductivity in each population of two crosses

The comparisons of relative conductivity in non-segregating generations, including 2 parents and F_1 of the 2 crosses were listed in table 4.11. The relative conductivity of 2 parents was extremely significant different with each other ($P<0.01$) and the means of relative conductivity of H077 and H113 were 45.63% and 40.85%, respectively. The means of relative conductivity of F_1 population for two crosses were 36.88% and 36.82%, respectively. Obviously, they both had a tendency toward the parent H113. The difference of relative conductivity between two F_1 of H077×H113 and H113×H077 was not significant, which means there was no significant cytoplasm effect on cryotolerance in the reciprocal crosses (表4.11).

表 4.11　Comparision of relative conductivity in non-segregating generations

Generation	Materials	Mean (%)	Range (%)	Standard error	Kurtosis	Skewness	t	P (two-tailed)
Parents	H077	45.63	43.73~48.79	0.39	-0.19	0.59	9.81**	0.00
	H113	40.85	38.74~42.90	0.29	-0.39	-0.12		
F_1	H077×H113	36.88	29.84~42.17	0.51	0.30	-0.48	0.08	0.94
	H113×H077	36.82	31.21~42.26	0.59	-0.46	0.03		

Note：**presents significant difference at 0.01 level.
https://doi.org/10.1016/S2095-3119 (12) 60040-9

2. 阅读图 4.16 和图 4.17，并展示您对图中所示数据的理解

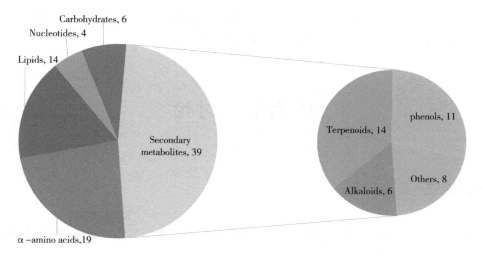

图 4.16　Compound pie chart of primary and secondary metabolites based on 82 significant metabolites.

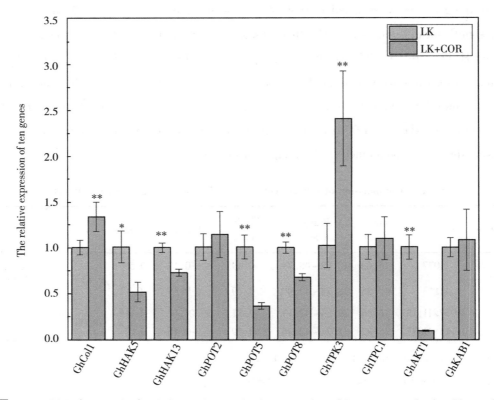

图 4.17 COR（coronatine）-induced changes in the expression of key genes conferring K transfer.

Note：For a given gene, one asterisk above the bars indicates a significant difference between the LK and LK+COR treatments（$P < 0.05$）, and two asterisks indicate a highly significant difference between the LK and LK+COR treatments（$P < 0.01$）according to the t-test. Bars are means ± SE, $n = 3$.

第六节 讨论

一、概述

讨论部分主要对研究结果进行解读、分析和讨论，并回答引言部分提出的问题。在本部分中，读者可以找到以下信息：

（一）通过总结研究的重要发现来回答所提出的问题

（二）研究结果的理由和原因，以及结果如何支持答案的解释，以及答案如何符合

关于该主题的现有知识

（三）通过提供研究者的解释和意见，指出研究的影响

（四）研究结果与现有研究结果的关联和比较

（五）研究的局限性和对进一步研究的建议。有时，结果部分和讨论部分可以合并在一起

二、特点

讨论部分通常遵循以下规则：

（一）讨论是从具体到一般的组织

在给出具体数据后，概括性地总结或得出结论。

（二）讨论的观点与引言部分的观点相同

作者在引言部分阐述了自己的观点，通常在这一部分保持观点的一致性。

（三）讨论部分中既有事实，也有观点或评论

作者不仅描述了有关结果的事实，而且提出了他或她对结果的意见或评论。

三、语步

在阅读本部分时，您应该关注以下3个主要语步：

（一）回顾本研究

这一举动提供了一个研究的目的和主要方法特点方面的研究概述。

（二）巩固成果

在这一举措中，通常要执行四个步骤，以履行一系列职能，详情如下。

第1步：报告结果

为了强调研究的结果，作者通常首先介绍研究结果，以便以后讨论。

第2步：解释研究结果的意义以及为什么其很重要

随后，作者给出了研究结果的原因，并解释了研究结果如何支持答案，以及答案如

何与现有的知识相适应。读了这一部分后，读者可能会明白这篇研究型论文的意义。作者通常以最重要的发现开始这一部分。

第3步：将研究结果与同类研究相关联

接下来，作者将研究结果与同类研究的结果联系起来，指出作者的研究与同类研究有何不同。因此，这项研究可以得到巩固。

第4步：为结果提供其他解释

研究的目的是发现一些东西。因此，作者经常仔细考虑对研究结果的所有可能的解释，而不是只关注那些符合作者先前假设的解释。

（三）陈述局限性和未来可能的研究方向

在这一举动中，作者分析了研究的局限性，并给出进一步研究的建议。

四、阅读技巧

（一）区分事实和观点

在讨论部分中，既有关于结果的事实，也有作者对结果的意见，所以在阅读这一部分时，有必要知道如何区分事实和意见。事实是对事情真相的陈述。或者情况真的是这样，具有客观性。意见是对一个主题的信念、判断或感觉的陈述，具有主观性。事实代表真实的东西，而意见只是信仰。事实可以用证据来证明。而意见必须有事实的支持，才能听起来合理。观点是有争议的，而事实通常是没有争议的。事实包括：

（1）数字、日期、年龄和表明事实信息的统计数据。

（2）已被证明的科学和历史信息。

（3）关于一个人、一个地方或一件事的具体信息。

意见通常以下列线索词和短语呈现：

（1）情态形容词，如 possible、probable、indefinite、unlikely 和 normal。

（2）表示程度的形容词，如 great、advanced、well-addressed、best、greatest 和 worst。

（3）表示频率和程度的副词，如 often、frequently、occasionally、usually、approximately、roughly、somewhat、always、never、all 和 sometimes。

（4）情态动词，如 can、could、may、ought 和 should。

（5）试探性动词，如 suggest、indicate、assume、estimate、tend、appear、seem 和 interpret。

（6）介绍性短语，如：to our knowledge, it is our view that... it can be argued that... it can thus be concluded that... 以及 one can assume that...

（二）识别惯用语

有一些惯用语可以用来识别研究结果的理由和原因，与已有研究结果的联系和比较，以及作者的态度和建议。

1. 用于指示研究结果的理由和原因的惯用语

（1）contribute to

（2）because of

（3）result in

（4）as a result of

（5）... could be responsible for...

（6）The results may be due to/caused by/attributed to...

（7）This is probably a consequence of...

（8）It seems that... can account for/interpret...

2. 用于关联研究结果与现有研究比较的惯用语

（1）... correlation was found between A and B.

（2）There was a (n) association between A and B.

（3）... did not demonstrate an association between A and B.

（4）A was highly associated with B.

（5）A was strongly associated with/inversely correlated with/in direct proportion to B.

（6）In contrast, there were significant relationships between A and B.

（7）No relationship was noted between A and B.

（8）The analysis showed no relationship between A and B.

（9）The results of... in the present study were in accordance with...

（10）These findings are in agreement/line/accordance with...

3. 用于指示作者态度、建议等的惯用语

（1）Based on the present data, we hypothesize that...

（2）These findings lend support to the hypothesis that...

（3）Based on these results, we propose that...

（4）It is proposed/recommended that...

(5) The authors propose/recommend that...
(6) It is suspected that...
(7) The authors' opinion is that...
(8) We feel/think/believe/consider that...

五、举例分析

(1) The mixed genetic model of major gene and polygene provided an efficient way to analyze the inheritance of complicated traits. (2) It had been used for studying some characteristics in cotton, i.e., yield, quality, earliness and plant architecture (Dong et al., 2010; Li et al., 2010; Yuan et al., 2002; Zhang et al., 2006). (3) These studies indicated that major genes of most of the important agronomic and economic characters in cotton were predicted, which supplied a theoretical evidence for guiding the breeding in cotton. (4) At present, the mixed model of major gene and polygene had been used for studying cryotolerance in some plants excluding cotton, and most of the reports were related with rice. (5) In this study, six populations including P_1, P_2, F_1, B_1, B_2 and F_2 from each of the two reciprocal crosses H077×H113 and H113×H077 were made up for studying the inheritance of cryotolerance in cotton during the overwintering period by using the mixed genetic model of major gene and polygene. (6) The results showed that the cryotolerance in cotton during the overwintering period was controlled by two additive major genes and additive-dominance polygene, which were basically consistent with the cryotolerance at the booting stage in rice (Dai et al., 1999; Yang et al., 2007), cucumber (Yan et al., 2009) and *Zoysia* grass (Guo et al., 2009). (7) In addition, some studies showed that the cryotolerance in some plants was controlled by one pair of major gene (Shen et al., 2004). (8) It indicated that the cryotolerance was a complicated quantitative trait whose inheritance was possibly related to some nonsystematic error factors in research such as plant materials, model quantities and generation population etc. (9) In this study, two additive major genes of cryotolerance in cotton during the overwintering period were predicted, which provided theoretical references on main-effect QTL identification using molecular marker technology.

语步1：句子（1）~（4）提供了关于研究方法主要特点的研究概述。

语步2：句子（5）~（9）巩固了关于主要研究结果并解释了研究的意义。

(10) Understanding the major gene and polygene genetic rule of quantitative traits, we would choose the appropriate means to enhance breeding efficiency. (11) Single cross recombination or simple backcross should be adopted to transfer positive major genes for quantitative traits, which belonged to typical major gene inheritance or mainly controlled by major gene; polymerization backcross or recurrent selection should be adopted to cumulate positive polygene for the traits, which belonged to typical polygene inheritance or were mainly controlled by polygene; the relative effects of major genes and polygene should be considered separately for the traits, which were controlled by major gene and polygene altogether (Gai et al., 2003). (12) In this study, the mean heritabilities of the major genes in B_1, B_2 and F_2 for the reciprocal crosses of H077×H113 and H113×H077 were 83.67% and 73.26%, respectively, and the mean heritabilities of polygene were 2.59% and 2.17%, respectively. (13) It was found that the heritability of the major gene was always higher than that of polygene in two crosses. (14) Therefore, in accordance with the breeding strategy of qualitative character, using the method of single cross recombination or simple backcross to transfer major genes of cryotolerance in cotton during the overwintering period would be suitable. (15) It was also found that the whole heritability (major gene heritability and polygene heritability) in F_2 was always higher than that in B_1 and B_2 in each of the crosses, indicating that selection in F_2 would be more efficient than other generations for cotton cryotolerance breeding.

语步3：句子（10）~（15）给出了进一步研究的建议。

 阅读练习

阅读下面关于一篇研究型论文的讨论，试着回答表4.12中的问题。

Effects of rootstocks on cryotolerance and overwintering survivorship of genic male sterile lines in upland cotton (*Gossypium hirsutum* L.)

(1) Chilling stress occurs at temperatures lower than the plant's normal growth temperatures but not low enough to cause ice formation. (2) Chilling is damaging for some plants pri-

marily because of membrane leakiness caused by inability to increase membrane fluidity to accommodate the lower temperature (Paul et al., 2006). (3) It has long been established that injured cells are unable to maintain the chemical composition of their contents and may, therefore, release electrolytes through damaged membranes. (4) The mechanism by which such events occur has still to be resolved; damage may be caused by reductions in cell volume following dehydration or by increasing concentrations of electrolytes (Deans et al., 1995). (5) The electrolyte leakage method of hypothermal injury assessment first described by Dexter et al. (1932) and the method of relative conductivity derived from it, Wilner (1960) provided a less subjective method for scoring hypothermal injury and it has been widely used (Eugénia et al., 2003). (6) The accumulation of compatible solutes in the cytoplasm, such as soluble sugar and soluble protein, contribute to cryotolerance by reducing the rate and extent of cellular dehydration, by sequestering toxic ions, and/or by protecting macromolecules against dehydration-induced denaturation (Steponkus, 1984). (7) Free proline has been proposed to act as a hydroxyl radical and singlet oxygen scavenger (Wang et al., 2007), and to alleviate free-radical damage induced by chilling stress.

(8) Grafting is regarded as a promising tool to broaden the temperature optimum of plant, the grafting process itself had no obvious effect on plant cryotolerance. (9) The increased cryotolerance of grafted plant was due to "the use of cryotolerant rootstock", plants grafted onto different rootstocks respond more or less differently to chilling (Jan et al., 2008). (10) This study concludes that grafting is an effective way to improve cryotolerance of cotton for the contents of soluble sugar, soluble protein and free proline in the bark tissue of the cryosensitive scion was increased after grafted onto the cryotolerant rootstocks which were selected from a chilling stress trial. (11) These changes are correspondence with cryotolerance in plant tissue; however, it is hard to discern which is critical. (12) Therefore, we supposed that acquisition of cryotolerance is a multifactor result of all these events and process (Wen et al., 2009). (13) That is why the subordinate function method in fuzzy mathematics was used to synthesize various physiological parameters correlated with cryotolerance for estimating accurately and comprehensively (Chen et al., 2007).

(14) GMS lines of plant could be used as sterile line and maintainer in breeding, and they have the conspicuous characteristics of fertility is easy to regain but hard to maintain by sexual reproduction. (15) In the present study, in order to maintain the fertility of GMS cotton by means of its perennial growth on the basis of frostless winters in Nanning, Guangxi autono-

mous region, GMS line A4 was grafted onto 7 different rootstocks (F118, F697, F098, F112, F113, P098 and P113), and the cryotolerance and overwintering survivorship of the grafted plants were investigated. (16) In this study, the physiological parameters of cryotolerance could be used for forecasting the overwintering survivorship, and the relative conductivity may be the first physiological parameter for forecasting cryotolerance or overwintering survivorship. (17) The results indicated that the cryotolerance and overwintering survivorship of GMS cotton could be improved by grafting, which is basically consistent with the researches in some plant species like cucumber (Ahn et al., 1999; Li et al., 2008), tomato (Jan et al., 2008) and grape (Zhang et al., 2009), and F113 appeared to be a valuable rootstock. (18) The cryotolerance of grafted plant depends on not only the cryotolerance of rootstock, but also the compatibility between scion and rootstock. (19) F113 was F_1 hybrid of an upland cotton crossed with island cotton P113, the cryotolerance of grafted plant A4/F113 was better than A4/P113 which implied that maybe the heterosis of cryotolerance in F113 was positive and its compatibility with upland cotton A4 was better than P113.

(20) In grafted plants, the rootstock absorbs water and nutrients, and composes hormone, proteins and metabolites, etc, which are transported to scion by graft union and affect the growth and development of scion. (21) Grafting is also a well established technique for the growth and production of cotton, such as for controlling cotton *Verticillium* wilt, increasing yield in continuous cropping cotton field (Hao et al., 2010) and alleviating leaf senescence of early-senescent scion when grafted onto late-senescent rootstock (Dong et al., 2008), etc. (22) The heterosis of hybrid is caused by the genetic composition of their parents which is not influenced by grafting, so there is no significant difference between hybrids produced from grafted GMS cotton and un-grafted GMS cotton.

(23) GMS upland cotton propagated by grafting once overwintered safely could grow for producing hybrid seeds about 3 years in tropical and partial subtropical regions (Zhang et al., 2010; Chen et al., 2010), and it would omit the matched maintainer of the GMS line. (24) Although it would need lots of labors to graft in the first year, but in the next two years, it would not need to plough, sow seeds, graft, etc. (25) It would save lots of labors, seeds, fertilizer, etc, and it is conducive to the protection of the environment as well. (26) However, it is still tedious in actual application, especially in the first year. (27) In the future, it would not be necessary to propagate male sterile plants by grafting if the perennial GMS line of cotton with high overwintering survivorship, good synthetic properties, and high combining a-

bility were bred. (28) However, it is very difficult to be achieved for the perennial cottons have two fatal weaknesses: the growth period is too long and the yield is too low, even the perennial GMS lines were bred, if the sterile plants of them were not grafted, the fertile plants should be removed before producing hybrids in the first year. (29) Consequently, grafting annual GMS scion onto perennial rootstock would be potential for perennial heterosis utilization in cotton, which could combine their merits and produce more F_1 hybrid cotton seeds with high quality and inexpensive.

表 4.12　研究型论文讨论部分阅读练习

问题	对应的句子
（1）哪些句子解释了结果是如何支持答案的？	
（2）哪些句子表明答案如何与该主题的现有知识相吻合？	
（3）哪些句子说明了解释和意见？	
（4）哪些句子解释了研究结果的含义？	
（5）哪些句子对未来的研究提出了建议？	

第七节　结论

一、概述

研究型论文的结论是在研究后对试验结果或者调查分析结果进行综合总结，提供了作者对所研究问题的回答、得出的结论和建议、研究成果的解释以及未来研究的方向和前景。

通常，这一部分包含以下主要信息：

（一）重申研究问题和论点

（二）对研究目的、问题及假设的回答和总结，即结论要从定量和定性两个角度对研究结果作出概括

（三）阐述研究结果对原有理论的验证、修正和拓展，以及其对相关领域或实践有何启示和影响

（四）强调研究的贡献和局限性，并对未来研究提出建议

二、类型

（一）以论点为导向的结论

以论点为导向的结论是指根据研究问题的答案、研究结果和论点，对整篇文章进行总结评价的结论部分，强调整篇文章的主题和中心论点，重申研究问题并陈述已提出的解决方案和建议。在这个结论部分中，作者应该总结文章中的主要观点和证据，以及说明这些观点和证据如何支持他们的主张。此外，作者还可以为未来的研究提供一些建议，并概述他们对所研究领域的贡献。其典型结构包括如下内容。

1. 介绍性陈述

重申研究的目的、问题或假设，以及开展的工作。

2. 回答研究问题

指出研究问题的答案，并强调研究的核心发现，以便在结尾处再次重申和强调。

3. 总结研究结果或发现

概述您在前面各部分中得出的重要结论或发现，并反映您对这些结果的理解。

4. 提出未来的研究方向和建议

此步骤可能需要评估当前研究的局限性，并用未来的方向和方法来探讨推进研究工作的方式。

5. 突出对该研究领域的贡献

探讨该研究如何填补所在领域的不足之处或开创新的研究方向，即强调该研究的独特性和创新性。

示例：

In this study, proteomic analysis shows the physiological and biochemical pathways of cotton plants responding to low-temperature stress and grafting. Chilling stressed carbohydrate, lipid, nucleotide, and amino acid metabolism, with relatively more active path-ways related to protein folding and degradation and activation of antioxidant enzyme system. However, oxidative phosphorylation, photosynthesis, carbon fixation, and other energy synthesis using light were weakened and material transport was slowed down. This indicates that the physiological and biochemical changes in cotton under low-temperature stress are very complex, involving multiple factors. Grafted plants improved the ability of scions to resist this stress, but the effect was not obvious in rootstock. This experiment lays the foundation for further research on differential proteins related to chilling stress or grafting, and the discovery of new cold tolerance genes.

(二) 面向领域的结论

这种结论是聚焦于总结研究领域内的问题、挑战、发展趋势等，而不是针对某个具体研究主题进行总结，包括对研究领域的总结和评价，讨论当前领域面临的困境和挑战，介绍新的研究进展、技术或方法，并提出未来研究的方向和需求。

示例：

The current study is based on the concept of genomic hybrid breeding, previously utilized in rice (Xu et al., 2014), which exploited the strategy of genome sequencing. The sequence data was then deployed to evaluate F_1 progenies' performance in hybrid breeding. An earlier study on rice revealed the power of SNP-directed yield estimation of F_1 hybrids. In the current study, 298 QTNs were uncovered in association with fiber quality as well as agronomic traits. A set of 271 hQTNs were detected with 19 highly stable heterotic loci in relation with LP, BW, FL, FS, and MIC based on their detection from five evaluated types of heterosis and/or four F_1 hybrid sets across a wide spectrum of environments. These discovered hQTNs and putative candidate genes related to HETEROSIS of quoted traits could be used further deliberately in marker-assisted breeding of forthcoming cotton hybrid breeding programs. Once the genotype-based predictions achieve relatively high levels of accuracy, the labor and time costs of hybrid breeding are greatly reduced. The reported information derived in this study is of practical and scientific significance for both cotton breeders and biologists engaged in elucidating the heterosis mechanism of fiber as it could assist in successful accomplishments in both domains.

三、语言特征

在结论部分，通常使用不同的时态。一般现在时表示总体结论；一般过去时表示本研究的初步结果和局限性；一般将来时表示未来的工作。

此外，结论部分还经常使用一些情态动词。例如，may 可以表示推论、含义和可能的应用，should、could 和 would 可以用来表达作者的评价、意见和建议。

在这一部分中，有一些惯用语表达不同的信息，主要有以下几个方面。

（一）表示总结的语言信号（表 4.13）

表 4.13　表示总结的信号词

信号词	信号词	信号词
accordingly	in conclusion	overall
as a consequence	in a word	therefore
as a result	in short	thus
consequently	in sum	to conclude
finally	in summary	to summarize
hence	on the whole	to sum up

（1）In conclusion, our findings provide strong support for the argument that...

（2）Overall, our study demonstrates that...

（3）These results clearly indicate that...

（4）It is evident from our findings that...

（5）Taken together, the results of this study provide strong evidence for the assertion that...

（6）Our findings strongly suggest that...

（7）Thus, it can be concluded that...

（8）Based on our study, we advocate for the idea/concept that...

（二）用于重述假设的惯用语

（1）As hypothesized/predicted/expected, the data supported our theory...

（2）Our findings support the hypothesis/claim that...

（3）Our results confirm/in line with the initial hypothesis/expectations that...

（4）The results of our study are consistent with the hypothesis that...

（5）Our study/findings provide support for the hypothesis that...

（6）These findings lend support to the hypothesis that...

（7）It appears that our hypothesis was valid/invalid.

（8）The data is consistent with our theoretical framework.

（9）In accordance with prior research...

（10）Our results confirm earlier studies...

(11) The results are in agreement with our initial hypothesis.

(三) 用于暗示影响的惯用语

(1) These findings have important implications...

(2) Our study has several practical implications...

(3) This research points to the need for further investigation in the field.

(4) The results of this study may contribute to the development of...

(5) Our research has the potential to inform policy-makers.

(6) This study opens the door for further exploration into...

(7) The implications of this study extend beyond...

(四) 用于反思局限的惯用语

(1) The present study is not without its limitations...

(2) One limitation of our study is the small sample size.

(3) Future research should aim to address these limitations.

(4) Despite these limitations, our findings contribute to the field in several important ways.

(5) It is important to acknowledge the limitations of this study.

(6) In retrospect, we could have addressed certain limitations through alternative methods.

(五) 用于表示未来前景的惯用语

(1) Future research could/should explore...

(2) Further research may investigate...

(3) This study provides a foundation for future research into...

(4) The current findings suggest several avenues for future research.

(5) The results of this study warrant further investigation.

(6) There is a need for future studies that address...

(7) To address these issues, it is important to...

四、举例分析

(一) 举例1

(1) This study elucidated morphological and physiological changes and potential plant defense ability induced by COR under LK from the perspectives of xylem sap proteomics. (2) In detail, COR significantly increased the average root diameter and the number of lateral roots but shortened root length in cotton seedlings, which were related to up-regulation of cobalamin-independent methionine synthase family proteins, auxin-responsive protein, and cell wall remodelling proteins such as dirigent-like protein, laccase, and the pectin lyase; it significantly lessened xylem sap volume and some cations contents, probably having connections with down-regulation of uclacyanin 1 and up-regulation of calmodulin-domain protein kinase 7; it significantly reduced the MDA content, mightly being associated with many PCD-related proteins down-regulation or loss. (3) Moreover, COR potentially weakened plant defense, owing to lessening or disappearing lipid-transfer proteins, signaling proteins, and so on. (4) This study fills a key gap in our knowledge of how COR impacts cotton root morphology and physiology and plant defense under K deficiency conditions in the xylem sap proteome's point of view.

句子(1)重申了研究的问题。

句子(2)和(3)总结了研究的主要结果并加以讨论。

句子(4)点明了研究的创新性和意义所在。

(二) 举例2

(1) Highly significant 46 microsatellites were discovered in association with FUI, LP, FS, FL, BW, MIC, FE, PH and FU. (2) Two-thirds of these significantly associated loci were scattered on D sub-genome, especially those of related to FS, FL and FU. (3) Also the pleiotropic effects of NAU2631, CM45 and GH501 loci on FUI, FS, FL and FE were detected. (4) A set of 96 exclusively

句子(1)~(4)总结了研究的主要结果。

favorable alleles were discovered primarily associated with BW, FL, FE and MIC mainly harbored by F_1s from C tester (A971 Bt). (5) To grab prominent improvement in mentioned influenced fiber quality and yield traits, we suggest the A971 Bt cotton cultivar as fundamental element in succeeding AM population development procedure to eliminate deleterious alleles residing at corresponding loci of superior alleles. (6) The output of this study can be helpful for plant breeders and researchers working to improve the yield and quality attributes of cotton for the efficient utilization of hybrid vigor.

句子（5）提出了对未来研究的建议。

句子（6）点明了研究的意义。

 阅读练习

1. 阅读以下两个结论，确定它们是以论点为导向，还是以领域为导向

Conclusion A

Hanging labels from the petioles (HLFP) decreased plant height, the number of fruiting positions and boll weight in four cotton genotypes. However, different genotypes seemed to respond differently to HLFP with respect to variables such as plant height and seed cotton weight. In addition, some cotton plants may exhibit delayed responses in some agronomy traits to HLFP treatment.

Conclusion B

In summary, this study identified 455 proteins from the xylem sap of field-grown cotton plants. These proteins may have multiple functions during cotton growth and development. They may also play an important role in cotton response to biotic and abiotic environmental stresses. The discovery of secreted proteins and proteins involved in transport provide useful insight into the communication mechanisms between cotton roots and the rest of the plant body.

2. 阅读以下结论，试着找出其中是否包含以下部分：研究问题的重述、结果的总结、结果的讨论、研究的意义、研究的局限性以及进一步的研究工作

K deficiency induced a series of changes of from ions uptake, organic substances metabolism, physiologies and morphologies in cotton. In details, K deficiency disturbed cations absorption and resulted in the acidity of xylem sap. Less K content decreased antioxidant capacity and led to membrane damage; made primary metabolism of sugar, protein and nucleic acid and secondary metabolism abnormal, lowering free sugar, water soluble protein, and polyphenol content, but enhancing amino acid content. Further insight into the results showed cotton presented positive adaption to K deficiency. For example, more osmosis regulating substances like sucrose and mannitol were produced to balance osmosis pressure reduction resulting from less K concentration; volatile terpenoids reduction mitigated the loss of carbon to maintain basic growth; nearly 30 folds increased glycerophosphocholine in xylem sap signaled leaf to coordinate leaf and root growth under K deficiency in comparison with sufficiency K.

第八节 致谢

一、概述

致谢部分是作者对于研究过程中得到帮助和支持的人、机构以及其他方面的感谢与表彰。在研究型论文中，致谢部分通常是在参考文献之前的一个完整的段落，反映了该论文所受到的智力、物力和财政支持。

致谢部分给读者留下了一个礼貌而支持性强的印象，表明作者对研究中协助者充满感激和尊重的心态，有助于内外交流和进一步研究的推动。一般来说，致谢部分从最重要的人开始，然后是基金会或研究团体获得的经济支持。

二、特点

研究型英语论文致谢部分需要符合礼貌、具体和简洁3个特点，这有助于表达感激之情的同时保持正式的学术风格，并使读者更容易理解作者对帮助者的感谢；同时，也是作者敞开胸怀拥抱身边支持者的一个重要途径。

（一）礼貌性

致谢部分需要作者以表达感激和尊重的心态，表达对研究中提供帮助和支持的个人或组织的感激之情。因此，致谢部分是客观的和正式的，应该注意用恰当的措辞，避免使用过于口语化或不当的词汇，以显示作者的礼貌和谦虚的态度。

（二）具体性

致谢部分应该对于所得到的帮助进行具体描述，使得读者能够理解作者所遇到的问题和解决方案，同时也能够了解所得到的帮助对于文章研究的重要性。

（三）简洁性

致谢部分语言应该简单明了，尽可能减少词汇用量，使用常用词汇，使读者一目了然。

三、阅读技巧

在致谢部分，有一些常用的句子结构。例如：

（一）The author (s) is/are indebted to... for their invaluable support throughout the research process

（二）The author (s) would like to express my sincere appreciation to... for their guidance, encouragement and constructive criticism

（三）The author (s) wish to extend my gratitude to... for their valuable insights and technical assistance

（四）This research would not have been possible without the generous help of...

（五）The author (s) would like to thank... for providing the funding support that made this research possible

（六）Without the assistance of... this project could not have been completed

（七）The author (s) is/are grateful to... for their encouragement, patience, and motivation throughout the challenging research period

（八）The author (s) would like to acknowledge the support of... in providing access to the necessary resources and facilities

（九）It is impossible to name everyone who provided assistance, but the author (s) would like to express my sincere appreciation to those who did

四、举例分析

（一）举例1

This work was supported by the Innovation Project of Guangxi Postgraduate Education,

China（2008105930901D015）.

（二）举例2

This project is partially supported by the State Science and Technology Support Project（2006BAD01A05 - xz02）and the Innovation Project of Guangxi Postgraduate Education（2008105930901D015）of China. The funders had no role in study design, data collection and analysis, decision to publish, or preparation of the manuscript.

第九节 参考文献

一、概述

参考文献是一种非常重要的正式引用方法，其作用主要有：

（一）突出自己的研究

通过引用前人的研究成果和论据，证明自己的研究已经比之前的学术成就进了一步。

（二）保证研究的可信度

科技论文的参考文献可以表现出作者的学术严谨性及根据准确性，这同时也保证了研究的整体可信度。

（三）为读者提供更深入的阅读素材

参考文献的使用，突出其他学者在论题上的重要贡献，同时也为读者提供了更多深入阅读和探究的资源。

二、特点

参考文献既是对前人研究成果的高度肯定，同时也为读者提供了更多深入阅读和探究的资源。其特点主要有以下几点：

（一）系统性

论文内容和引用的参考文献之间存在逻辑连贯性。

（二）客观性

参考文献应该具备客观性，遵循学术规范标准进行引用，不受主观因素的影响。

（三）细致性

参考文献应该将每个相应的原始信息完整地展示出来，以便读者可以快速获取并核对，避免错误引用。

（四）时效性

参考文献要尽量引用最新的文献，以保证研究的创新性。

三、类型

在研究型论文中，通常有两种类型的参考文献，即文内参考文献（也称为文内引用）和文末参考文献。

（一）文内引用

文内引用是指学术论文中引用的参考文献、引文、摘要或对另一来源（如书籍或期刊论文）的转述。可以是直接的也可以是间接的。直接引用是指一个字一个字地引用论文，并加上引号。例如：

Based on this, a breeding strategy called "planting in temperate regions and breeding in the tropics" was employed for rice (Liang et al., 2020), maize (Eagles and Lothrop, 1994), and sweet potato (Lu et al., 1989).

间接引用是指对原文进行转述或概括。例如：

In cotton, due to limited resources and negative cytoplasmic effects, CMS (Cytoplasmic Male-Sterile) lines have not been widely used (Zheng et al., 2019; Li et al., 2021).

Some wild cotton lines become sterile after transplanting from the origin (Zhang et al., 2022), showing traits such as non-flowering, non-dehiscence of anthers, and self-incompatibility.

（二）文末参考文献

文末参考文献，也被称为参考文献列表，是一个完整的参考文献列表，给出了学术论文中引用的所有来源的完整引用。

四、引文格式

参考文献的引用内容包括文章标题、作者、期刊名称、发表时间、页码、网址等等，为了使不同论文的参考文献保持一致，便于阅读，在引用时需按照规范格式完整地给出，尽可能清晰、简洁、恰当地说明信息，这种方式被称为引文格式。自然科学类的不同SCI期刊所要求的参考文献格式虽然千变万化，但基本上可以归结为两种参考文献格式：一种是"著者-出版年制"，一种是"顺序编码制"。

（一）著者-出版年制

著者-出版年制（Author-Year）起源于美国心理学协会（APA），是一种基于作者/年份的风格。在这种风格中，强调的是作者和作品的出版年份。

1. 著者-出版年制格式下的正文内引文

在著者-出版年制格式中，文本引用由作者的姓氏和出版年份组成。例如：

Coronatine (COR) could ameliorate the negative effects of abiotic stresses on plant performance (Xu et al., 2020; Zhou et al., 2015).

如果作者的名字出现在文本中，出版日期只会出现在括号中。例如：

The analysis software was provided by Dr. ZHANG Yuan-Ming, and the genetic parameters were estimated with the method, referred to Gai et al. (2003).

引用两位作者的作品时，括号中会显示两位作者的姓氏。在引文中使用"and"或符号"&"（不同的期刊规定的格式有所不同）来分隔人名，但在正文中使用"and"这个词。例如：

However, these resistance markers almost are secondary metabolites produced by plants that help fight against various stresses in nature (Tiwari and Rana, 2015).

After 24 hours, the barks of the branches were peeled and cut into chippings, then their relative conductivity were measured by using an electric conductance meter as described by Dionisio-Sese and Tobita (1998).

如果引用了 3 个或 3 个以上作者的同一作品，则在正文第一次和随后的引用中，在第一作者的姓后面加上"et al."。例如：

Although K has many positive effects on cotton yield and fiber quality, inadequate use of K is a challenging issue in cotton production under an intensive farming system (Shahzad et al., 2019).

Liu et al. (2014) identified 54 miRNAs including 47 conserved and 7 new miRNAs from island cotton using deep sequencing; their study also show that miR160, miR167, miR171, miR172 and miR827 were highly expressed in fiber initiation stage comparing to the elongation and secondary wall biosynthesis stage.

2. 文末参考文献著者-出版年制格式

如果引用了一本书中的一段文字，那么参考文献可以如下所示：

Choudhary B, Gaur K, 2010. Bt Cotton in India: A Country Profile [M]. ISAAA Series of Biotech Crop Profiles. ISAAA: Ithaca, NY.

如果引用的是期刊论文，那么参考文献可以如下所示：

Zhang X, Chen GP, Zhou RY, 2010. Effect of perennial cultivation on "Dong A" genic male sterile lines in annual upland cotton [J]. Guihaia, 30 (3): 391-394.

如果引用了某个网页的文字，则可以如下所示进行引用：

Pritchard JK, Wen X, Falush D (2010). Documentation for structure software version 2.3. University of Chicago. http://pritch.bsd.uchicago.edu/structure.html. Accessed 07 Sep 2014.

一般来说，在著者-出版年制格式中，所有的参考文献都是按第一作者姓氏的字母顺序排列的。当想找某一位作者的原文献时，只要按照字母顺序去找就可以了。

（二）顺序编码制

顺序编码制（Numeric）源于电气和电子工程师协会（IEEE）制定的标准参考文献样式，即基于广泛使用的芝加哥参考文献样式。在顺序编码制格式中，使用了一个编号系统来确保论文易于阅读。正文中的编号与研究型论文结尾处的编号相关联，以明确哪一个来源对论文的哪一部分做出了贡献。

具体来说，顺序编码制是按论文的正文部分（包括图、表及其说明）引用的文献首次出现的先后顺序连续编码，参考文献的序号均用阿拉伯数字标明。在正文内引用时将序号标注于有关词组或段落相应处的文句后方，引文写出原著者，序号标在著者的右上角，如×××（et al.）[1]；如未写著者姓名，序号应放在引文之后。书写时要注意对2篇以上不连续序号以逗号","隔开，比如用［1，5］，［1，3，7］的方式；对3篇以上连续的序号，只标注引用文献始末的编排序号，中间使用起止"-"连起，比如以［1-5］方式标注。

1. 顺序编码制文内引用

数字或括在方括号内的数字，比如［1］或［2，6］或［1-6］，放在正文中以表示相关的参考文献。引文是按照其在正文中出现的顺序编号的，每个引文对应于出现在末尾的编号引用。例如：

Cotton provides the primary textile in the world. Heterosis has been observed in many crops, including cotton, and it could be expressed over mid parent[1] or over a check hybrid[2,3] in cotton.

2. 顺序编码制文末格式

在顺序编码制的风格中，如果引用了一本书中的一段文字，那么文末引用的内容可以像下面这样。

［12］Hallauer AR, Carena MJ, Filho JBM. Quantitative genetics in maize breeding［M］. Iowa: State University Press, 2010.

如果引用学术期刊中的论文，那么参考文献可以如下所示。

［29］Xiong E H, Zhang C, Ye C X, Jiang Y H, Zhang Y L, Chen F, Dong G J, Zeng D L, Yu Y C, Wu L M. iTRAQ-based proteomic analysis provides insights into the molecular mechanisms of rice formyl tetrahydrofolate deformylase in salt response［J］. Planta, 2021, 254（4）: 76.

如果引用了某个网站上的论文或报告，那么引用的内容可以如下所示。

［8］Hsu HH, Gale F（2001）. Regional shifts in China's cotton production and use. Cotton and wool situation and outlook. Economic Research Service, USDA. http://www.ers.usda.gov/ContentPages/4093190.pdf. Accessed 15 Nov 2014.

值得注意的是，在顺序编码制样式中，所有的结束引用都是按编号顺序列出的，这与著者-出版年制样式截然不同。

五、阅读技巧

当阅读研究型论文的参考文献部分时，您需要了解所使用的不同风格的参考文献，并尝试识别不同类型的文献来源（无论是一本书，期刊论文，学位论文，或其他来源），然后试着识别每一个参考文献中包含的信息，从而找到信息的来源。阅读这些参考文献，可以让读者更深入地了解作者研究的背景和相关工作，从而更好地理解和评价作者的贡献和创新，同时也有助于扩展自己对研究领域的认识和了解。以下是阅读研究型论文的参考文献的一些技巧：

（一）注意参考文献格式

不同的期刊和出版物有不同的参考文献格式要求。在阅读文献时，需要注意其格式是否符合期刊或出版物的要求，以确定其可靠性和学术水平。

（二）查找原始文献

在读到参考文献中提到的某一篇文献时，应该尽量查找该文献的原始版本。这有助于更好地了解研究成果的细节，从而对论文提出更有建设性的意见和建议。

（三）分辨可靠性

在阅读参考文献时，需要对其可靠性进行评估。一些最受欢迎的出版物、期刊或重要会议的文章可能比其他杂志的文章更可靠。同时，需要对作者的资历、研究方法和数据来源等方面进行评估，从而得出对文章的可靠性和学术价值的评估。

（四）把握文献关系

阅读参考文献可以帮助了解当前研究领域的现状，同时也有助于发现相关研究和前沿方向。在阅读文献时，需要注意该文献与其他文献之间的关系，包括与其相关的和互相引用的文献等，以便更好地掌握当前研究领域的趋势和发展方向。

六、举例分析

（一）举例1

1. 正文内的引用

　　Plants are immobile and therefore unable to escape from threats or unfavorable environments. Consequently, they have evolved diverse mechanisms for coping with and mitigating the effects of various stresses[1]. The adverse effects of disadvantageous environmental conditions, such as excessively high or low temperatures, salt levels, and drought on plant growth and development have been extensively documented. However, the effects of mechanical stresses caused by factors such as wind, rain, physical contact, wounding, and gravity on plant growth and development have not been studied in such detail[2,3].

2. 文末参考文献

[1] Potters G, Pasternak TP, Guisez Y, Palme KJ, Jansen MAK. Stress induced morphogenic responses: growing out of trouble? [J]. Trends in Plant Science, 2007, 12: 98-105.

[2] Braam J. In touch: plant responses to mechanical stimuli [J]. New Phytologist, 2005, 165 (2): 373-389.

[3] Li ZG, Gong M. Regulation of calcium messenger system on mechanical stimulation-induced H_2O_2 burst in tobacco (*Nicotiana tabacum* L.) suspension culture cells [J]. Plant Physiology Communications, 2010, 46: 135-138. (In Chinese)

（二）举例2

1. 正文内的引用

The cell wall is composed of carbohydrate-like cellulose, lignin, and pectin, in addition to glycoproteins such as arabinogalactan proteins (AGPs). Root diameter thickening and lateral root formation are associated with a loosening of the cell wall and its structural remodeling (Wang et al., 2017; Zhang et al., 2016). The disease resistance-responsive (dirigent-like protein) family protein combines with laccase to form (+) pinoresinol, which mediates lignin synthesis (Burlat et al., 2001). Pectin can enable adjacent cell walls to adhere together, affecting tissue differentiation (Hongo et al., 2012). AGP is known to play a role in the deposition of pectin in the cell wall (Leszczuk et al., 2019).

2. 文末参考文献

Wang Y L, Meng Z G, Liang C Z, Meng Z H, Wang Y, Sun G Q, Zhu T, Cai Y P, Guo S D, Zhang R, Lin Y, 2017. Increased lateral root formation by CRISPR/Cas9-mediated editing of arginase genes in cotton [J]. Science China. Life Sciences, 60 (5): 524-527. https://doi.org/10.1007/s11427-017-9031-y.

Zhang Z Y, Chao M N, Wang S F, Bu J J, Tang J X, Li F, Wang Q L, Zhang B H,

2016. Proteome quantification of cotton xylem sap suggests the mechanisms of potassium-deficiency-induced changes in plant resistance to environmental stresses [J]. Scientific Reports, 6 (1): 21060. https://doi.org/10.1038/srep21060.

Burlat V, Mi K, Davin L B, Lewis N G, 2001. Dirigent proteins and dirigent sites in lignifying tissues [J]. Phytochemistry, 57 (6): 883-897. https://doi.org/10.1016/S0031-9422 (01) 00117-0.

Hongo S, Sato K, Yokoyama R, Nishitani K, 2012. Demethylesterification of the primary wall by *PECTIN METHYLESTERASE*35 provides mechanical support to the Arabidopsis stem [J]. Plant Cell, 24 (6): 2624-2634. https://doi.org/10.1105/TPC.112.099325.

Leszczuk A, Kozioł A, Szczuka E, Zdunek A, 2019. Analysis of AGP contribution to the dynamic assembly and mechanical properties of cell wall during pollen tube growth [J]. Plant Science, 281: 9-18. https://doi.org/10.1016/j.plantsci.2019.01.005.

第五章

阅读综述型论文

综述型论文是学者们经常提到的另一种主要类型的学术论文。综述型论文是一种针对某一主题或问题的综合性研究，其目的是全面梳理和总结当前领域内已有的相关研究成果和观点，以促进学术交流和思考。

与研究型论文往往侧重于某个研究主题的某个方面不同，综述型论文通常不是报道新的试验结果，而是对该领域最新研究成果的概述、归纳和综合分析，以及对未来研究方向和发展趋势的讨论。

综述型论文一般由以下几个主要部分组成：
（1）标题和作者单位。
（2）摘要。
（3）引言。
（4）正文。
（5）结论。
（6）参考文献。
（7）致谢。

考虑到综述型论文的作者署名和致谢的部分类似于研究型论文，本章将省略这部分内容。

第一节 综述型论文介绍

一、概述

综述型论文概括了某个领域的现状和发展，特别是突出了相关领域的新发展、新趋势、新发现、新技术。

在综述型论文中，创新性主要表现在以下几个方面：

（一）研究视角

综述型论文应该从全新的、前瞻性的视角出发，进行对已有研究的梳理、分析和总结，从而形成自己的研究思路和创新观点。

（二）问题解决

综述型论文应该能够针对当前学术领域中存在的问题，提出自己的创新性解决方案或者改进建议。

（三）研究方法

如果综述型论文的研究内容需要使用研究方法，那么该方法需要具备创新性，或者根据已有研究方法进行改进并取得新的研究成果。

（四）结果和贡献

综述型论文应该明确阐述自己的研究成果以及对学术领域的贡献，这些都需要体现出创新性。

（五）实践应用

若综述型论文研究成果能够被实践工作所应用，那么可以通过实践证明其创新性，更好地凸显该论文的价值。

综述型论文是一种系统性、综合性的学术论文，其主要目的是对某一领域或问题进行全面的梳理和总结，旨在汇总已有的研究成果与结论，提出未来的研究方向和重点。综述型论文通常会对当前某研究领域内的研究进展、发现和知识体系做一个系统的介绍和描述，以便更好地促进学术交流和进一步的研究。因此，一篇综述型论文应列出其中引用的支持性参考资料。

一篇综述型论文主要由标题和作者归属、摘要、引言、正文、结语、参考文献和致谢组成。通常首先介绍在该领域工作的历史进程、最近的重大进展和发现、技术的优缺点，然后提出研究中的重大差距，并在这一部分之后讨论当前的争论，最后提出本文将要讨论的一些想法。

二、类型

根据不同的标准，综述型论文可以分为不同的类型。

综述型论文可以根据其侧重点和方法学特点，分为传统综述和系统综述（又称系统评价、集成分析、综合分析）。传统综述通常侧重于对多个文献进行整合和概括，而不是像系统综述那样使用明确的方法学规范进行搜索、筛选和评价。根据数据类型，系统综述又可以分为定性系统综述和定量系统综述（即 Meta 分析，荟萃分析）。

根据目的和内容，综述型论文可以分为介绍性综述、发展性综述、结果性综述和争议性综述。介绍性综述只是提供说明性的事实，统计数据和意见，对原来的论文没有任何评论。发展性综述讨论某个地区或围绕某个主题的发展，重点关注其历史发展。结果性综述对相关文献进行分类，提炼文献要点，并对某些方面进行整理和讨论。争议性综述对某一研究领域内的有代表性的争论观点进行了总结和综合。

按时间顺序，综述型论文可分为回顾性综述和前瞻性综述。回顾性综述根据某一学科的历史发展，按顺序对所回顾的文献进行整理。前瞻性综述集中于某研究领域的现有知识和新发现。

根据作者的参与程度，综述型论文可分为归纳性综述和批评性综述。归纳性综述强调从被综述的内容中分析事实和观点。批判性综述对研究结果进行描述、分析和讨论，希望能产生批判性的评价和对综述内容的建设性解释。

根据综述内容的范围，综述型论文可以分为宏观综述和微观综述。宏观综述则较为全面，内容涵盖范围较广，如某个研究领域或学科的大方向。微观综述通常涵盖小范围的内容，但对某个主题的分析更为透彻。

此外，还有一些其他的综述型论文，如调研论文、论文评论、书评。调研论文，通常包含一个广泛的文献综述和许多关于科学调查的不同方面的技术细节。论文评论是以一篇论文为基础，重点是对论文进行评价。书评是对一本书的质量、影响和意义的描述、批判性分析和评价。

三、综述型论文和研究型论文的区别

综述型论文和研究型论文的区别在于，研究型论文是报告原始研究的方法和结果的主要来源。对原始数据进行收集和分析，并从分析结果中得出结论。综述型论文是关于其他出版物的第二来源，并不报道原创性研究。综述型论文提供了一个主题的现有文献的综述，还利用综述的文献来寻找新的研究方向，加强对现有理论的支持，和/或在现有的研究中的模式。

第二节 标题

一、概述

综述型论文的标题通常包含一些描述性词语，以反映其主要内容和焦点。例如，一篇系统综述的标题可能会涵盖以下元素：领域或话题名称、搜索策略、数据提取和分析方法、评估标准、研究结论等。一篇传统综述的标题则可能更为简短直白，旨在概括其主题或重点。通常，综述型论文的标题是简洁的，但在有一个很长的副标题的情况下，下列规范是必要的。

（一）使用冒号或破折号分隔主标题和副标题。冒号通常用于学术论文，而破折号通常用于新闻或杂志文章

（二）主标题应该简洁明了，突出论文的核心内容，不超过12个字

（三）副标题应该详细描述论文的内容，提供更多的细节和背景信息。可以包含关键词、研究对象、方法或主题等相关信息

（四）副标题的长度应该适中，不宜过长或过短。通常不超过20个字

二、阅读技巧

综述型论文的标题通常是信息性的，包括一些重要的术语，这些术语可以表明论文的性质和主题，因此根据其标题中的单词可以帮助理解论文中的信息。

综述型论文的标题经常包含 review 一词，如"A review of new advances in mechanism of graft compatibility-incompatibility"或"Forms of soil potassium：A review"。除了 review 一词，其他如 advances、meta analysis、overview、progress、revisit、survey、theory、thought 等词也可以作为判断论文是否为综述型论文的线索。因此，在阅读一篇综述型论文的标题时，你需要做的是通过寻找标题中的线索词来识别重点的综述内容。

（一）综述型论文标题中常用的线索词

表 5.1 所列是一些识别综述型论文标题中常见的线索词。

表 5.1 综述型论文标题中常见的线索词

线索词	线索词	线索词
achievement	discussion	progress
accomplishment	evaluation	promise
advance	frontier	pros and cons
advantage and disadvantage	framework	recent research
application	historical review	review
approach	implication	revisit
benefit	insight	significance
challenge	meta analysis	summary
characteristic	method	tendency
classification	outlook	theory
comparison and contrast	overview	thought
current status	potential value	trend
development	problem	view

（二） 常见的综述型论文标题

以下是一些常见的综述型论文标题：

（1） A comprehensive/systematic review/summary/overview/survey of … field/area

（2） An overview of developments in … area

（3） The latest progress in… research

（4） Trends/Frontiers in… field

（5） Current understanding of…

（6） Recent advances in …

（7） Historical perspectives on…

（8） New insights into…

（9） Emerging developments in…

三、举例分析

（一） 举例 1

A new perspective on Darwin's Pangenesis

在这个标题中，您可以看到 perspective 这个提示词，这有助于您确定这是一篇综述型论文,而综述的重点内容是关于"Darwin's Pangenesis(达尔文的泛生论)"的新观点。

(二)举例 2

Lysenko and Russian genetics:An alternative view

在这个标题中,线索词是 view。

 阅读练习

阅读下面的综述型论文标题,在线索词下面划横线。

(1) Current overview of allergens of plant pathogenesis related protein families

(2) Grafting in cotton:A mechanistic approach for stress tolerance and sustainable development

(3) New insights into plant graft hybridization

(4) The ratoon rice system with high yield and high efficiency in China:Progress, trend of theory and technology

(5) The influence of Darwin's pangenesis on later theories

第三节　摘要

一、概述

综述型论文的摘要通常出现在论文的开头,目的是对论文的主要内容进行概括,通常包括 5 个部分:

(一)研究背景

即本研究领域的发展现状、理论基础和存在的问题。

（二）研究目的

说明本综述的目标或目的。

（三）主要发现（内容）

综述的主要内容或概述。

（四）研究结论

指出研究的结果，并指出其对该领域或主题的意义和启示作用。

（五）未来研究

指出目前研究的差距和未来研究的方向。

此外，在系统综述摘要中，还要简要描述研究采用的方法、数据来源和分析技术等。

二、阅读技巧

（一）语言特征

综述型论文摘要部分通常具有简洁明了、逻辑清晰、使用关键词和惯用语等语言特征。按照上面提到的五个部分来阅读一篇综述型论文的摘要将是有益的。

通常，研究背景的信息是用现在时来描述的，带有一些时间状语，如"in the past years"和"in recent years"。

对于研究目的，可以以"here..."和"this paper..."为线索，一般使用现在时。

对于综述的细节，您可能会发现一些线索，如"specifically, we have described and discussed..."和"the discussion covers..."。一般用现在时来描述综述的细节。

结论通常用现在时态和将来时态来表述。例如"this review may help..." "we propose..."和"this technique will be extremely useful in helping..."可以帮助理解这一部分。

最后，作者经常使用现在时和过去时来表示目前研究的空白，现在时和将来时来表示未来的研究。例如："future perspectives on... are presented"和"the present review outlines the opportunities and limitations of..."等表述可以作为定位这部分的线索。

（二）词汇信号

综述型论文的摘要部分常用的词汇信号包括：

1. Review 或 systematic review

这些词汇表明论文是一篇对某个主题进行文献综述或系统综述的论文。

2. Objective

这个词说明论文有特定的目标或目的。

3. Analysis 或 synthesis

这些词汇表明作者对主题进行了分析或综合了现有研究。

4. Findings 或 conclusions

这些词汇强调了综述型论文的关键结果或结论。

5. Implications

这个词暗示论文讨论了研究结果对未来研究或实践的影响。

6. Limitations

这个词说明论文讨论了现有研究的局限性或不足之处。

7. Recommendations

这个词说明论文提供了未来研究或实践的建议。

这些词汇信号有助于引导读者理解综述型论文的目的、方法和结果，并提供一个清晰的全文概述。

（三）惯用语

摘要的每一部分都有一些惯用语，主要有以下几个惯用语：

1. 研究背景的惯用语

（1）Analysis of the literature suggests that...

（2）The area of... is gradually expanding cross the globe.

（3）With the ever-increasing need for...

（4）The study of... has become increasingly important due to...

（5）As the world becomes more globalized and interconnected, understanding... has become crucial.

（6）Notwithstanding substantial progress in... several fundamental challenges remain in the field.

（7）The existing literature on... has mainly focused on... leaving... largely unexplored.

（8）While some previous studies have examined... few have considered...

（9）The importance of... is underscored by its impact on...

（10）Given the significant role that... plays in... it is critical to better understand...

2. 研究目的的惯用语

（1）This review paper aims to...

（2）This paper presents a review of...

（3）This paper reviews an important topic within the broader framework of...

（4）A systematic/comprehensive/thorough review has been performed following...

（5）This review takes... as an example to summarize...

（6）This review describes/investigates...

（7）The purpose/aim/objective of this review is to...

（8）This paper provides a comprehensive overview of...

（9）This review seeks to achieve the following objectives...

（10）This literature review aims to provide a comprehensive analysis of...

3. 主要发现（内容）的惯用语

（1）This review examines...

（2）The review highlights...

（3）Several studies have reported...

（4）The literature suggests that...

（5）The review identifies...

（6）The analysis reveals...

（7）Our analysis reveals that...

（8）Several key conclusions can be drawn from this review...

（9）This meta-analysis confirms that...

（10）This review contributes to a better understanding of...

4. 结论的惯用语

（1）The review provides a comprehensive overview of...

（2）Overall, the review contributes to...

(3) The findings of this review reveal that...

(4) In conclusion, this review highlights...

(5) The review contributes to...

(6) In conclusion, this review...

(7) Based on the evidence presented in this review, it can be concluded that...

(8) It can be inferred from the research that...

(9) To sum up, this review...

(10) Thus, we can conclude that...

5. 未来研究的惯用语

(1) Additional investigation is necessary to...

(2) Areas for further exploration include...

(3) Futureresearches/studies could/should/need to investigate/ explore...

(4) Further investigation is needed in the field of...

(5) Future/further researches/studies could/should/need to focus on...

(6) The review highlights the need for...

(7) This study identifies a gap in the literature that could be addressed by future research.

(8) Potential avenues for future research include...

(9) In order to fully understand... additional research is necessary to...

(10) More research is warranted to determine...

三、举例分析

(一) 例1

(1) Ratoon is the stub or root of a perennial plant that is commonly retained after harvest to produce a following crop. (2) This paper presents a review of ratoon cotton in relation to a broader framework that has been examining perennialization of agriculture for the benefit of ecology and economy. (3) Cotton is botanically indeterminate, but has been treated as an annual after domestication, yet the habit of perenniality is retained and the plants begin to resprout after	句子(1)和(2)说明了研究目的。 句子(3)~(6)说明了研究背景。

the first harvest. (4) In some cropping systems, this tendency is exploited using the "ratooning" practice (i.e., growing one or more crops on the rootstock of the first). (5) Ratooning has declined for various reasons such as an increase in the prevalence of pests and diseases and overwintering risk. (6) However, ratooning has many benefits such as no annual tillage before sowing, a well-established root system, and high yield. (7) The three methods of ratooning offer flexibility to balance the environmental and economic benefits in agriculture. (8) The greatest environmental benefits arise from perennial ratoon cropping of semi-wild cotton, and the greatest economic benefit is obtained from biannually cropping modern annual cultivars. (9) However, an optimum solution would be provided by perennial cropping annual cultivars. (10) To realize both environmental and economic benefits, research is needed in the following main areas: preventing the buildup of pests and diseases, breeding the most suitable cotton cultivars for ratooning, and developing light and simplified cultivation (LSC) systems for ratoon cultivation.

句子(7)~(9)描述了主要发现。

第(10)句指出了该领域未来的研究方向。

(二) 例2

(1) The area of salinized land is gradually expanding cross the globe. (2) Salt stress seriously reduces the yield and quality of crops and endangers food supply to meet the demand of the increased population. (3) The mechanisms underlying nano-enabled plant tolerance were discussed, including ①maintaining ROS homeostasis, ② improving plant's ability to exclude Na^+ and to retain K^+, ③ improving the production of nitric oxide, ④increasing α-amylase activities to increase soluble sugar content, and ⑤decreasing lipoxygenase activities to reduce membrane oxidative damage. (4) The possible commonly employed mechanisms such as alleviating oxidative stress

句子(1)和(2)说明了研究背景。

句子(3)~(5)说明了研究内容。

damage and maintaining ion homeostasis were highlighted. (5) Further, the possible role of phytohormones and the molecular mechanisms in nano-enabled plant salt tolerance were discussed. (6) Overall, this review paper aims to help the researchers from different field such as plant science and nanoscience to better understand possible new approaches to address salinity issues in agriculture.

第(6)句说明了研究目的。

 阅读练习

阅读以下综述型论文的摘要，并尝试识别它们的结构并完成表5.2和表5.3的练习任务。

摘要1

Cotton production is challenged by high costs with multiple management and material inputs including seed, pesticide, and fertilizer application. The production costs can bedecreased and profits can be increased by developing efficient crop management strategies, including perennial cotton ratoon cultivation. This review focuses on the role of ratoon cultivation in cotton productivity and breeding. In areas that are frost-free throughout the year, when the soil temperature is suitable for cotton growth in spring, the buds of survived plants begin to sprout, and so their flowering and fruiting periods are approximately 4-6 weeks earlier than those of sown cotton. Due to the absence of frost damage, the ratoon cotton continues to grow, and the renewed plants can offer a higher yield than cotton sown in the following season. Moreover, ratoon cultivation from the last crop without sowing can help conserve seeds, reduce labor inputs, and reduce soil and water loss. In this review, the preservation of perennial cotton germplasm resources, the classification and genome assignment of perennial species in the cotton gene pools, and effective strategies for the collection, preservation, identification, and utilization of perennial cotton germplasms are discussed. Ratoon cultivation is the main driver of cotton production and breeding, especially to maintain male sterility for the utilization and fixation of heterosis. Ratoon cultivation of cotton is worth adopting because it has succeeded in Brazil, China, and India. Therefore, taking advantages of the warm environment to exploit the indeterminant growth habit of perennial cotton for breeding would be an efficiency-increasing, cost-saving, and eco-friendly approach in frost-free regions. In the future, more attention should be given to ratooning perennial cotton for breeding male-sterile lines.

表 5.2　综述型论文摘要阅读方法练习 1

综述摘要的 5 个部分	对应的句子序号
（1）研究背景	
（2）研究目的	
（3）主要发现（内容）	
（4）研究结论	
（5）未来研究	

摘要 2

This paper reviews an important topic within the broader framework of the use of ratoon cotton for the development of a cost-saving and efficient method for the perennial production of hybrid cotton seeds. Cotton has a botanically indeterminate perennial growth habit and originated in the tropics. However, cotton has been domesticated as an annual crop in temperate areas worldwide. Ratoon cultivation has important application value and is important for cotton production, breeding and basic research. In particular, ratooned male-sterile lines have four advantages: an established root system, an indeterminate flowering habit, ratooning ability and perennial maintenance of sterility in the absence of a matched maintainer. These advantages can help reduce the costs of producing F_1 hybrid cotton seeds and can help breed high-yielding hybrid combinations because ratooning is a type of asexual reproduction that allows genotypes to remain unchanged. However, ratooning of cotton is highly complex and leads to problems, such as the accumulation of pests and diseases, decreased boll size, stand loss during severe winters and harmful regrowth during mild winters, which need to be resolved. In summary, ratoon cotton has advantages and disadvantages for the production of hybrid cotton seeds, and future prospects of ratooning annual cotton for the perennial utilization of heterosis are promising if the mechanization of seed production can be widely applied in practice.

表 5.3　综述型论文摘要阅读方法练习 2

综述摘要的 5 个部分	对应的句子序号
（1）研究背景	
（2）研究目的	

（续表）

综述摘要的 5 个部分	对应的句子序号
（3）主要发现（内容）	
（4）研究结论	
（5）未来研究	

第四节　引言

一、概述

综述型论文的引言部分解释了综述的基本原理和目的。一篇综述型论文的引言通常包括研究背景、目的和范围、先前的文献评价、综述的内容及其意义等部分。此外，在引言部分，还会引入与本研究相关的关键词或术语，以便读者理解本研究主旨和内容。例如，"In this paper, we use the term 'X' to refer to..."。

综述的引言通常包括以下 4 个语步：

（一）研究背景

通过陈述主题、现状等来建立研究领域，即明确本研究的目的和范围，这有助于建立对本研究主题的基本认识。

（二）以前的研究

通常总结以往的研究，重点是综述的理由。

（三）综述的内容

通常介绍本研究的组织结构和各部分内容，概括出每个部分所包含的主要内容和目标，侧重于研究主题的内容。

（四）意义和/或对未来研究的建议

强调了本研究的意义，指出了本研究的不足之处，并提出了今后的研究方向。

需要注意的是，不同类型的综述型论文可能会在引言部分中强调不同的方面。例如，系统综述的引言部分可能更侧重于解释为什么该问题或话题很重要，并概述先前研究的方法和结果，而理论综述的引言部分则可能更侧重于阐述当前领域的理论背景和对该领域的贡献。

二、阅读技巧

（一）语言特征

在阅读一篇综述型论文的引言部分时，请把注意力集中在上面提到的 4 个语步上，这可能有助于您更好地理解这一部分。需要强调的是，阅读引言的过程需要注意整体理解引言，把握论文的主题和目标，同时要关注文献质量和作者的论述框架。以下是各个语步的一些语言特征。

1. 研究背景

在这一语步中，强调了相关研究的重要性、相关研究的现状、某些技术的应用等。因此，这个语步中的动词通常用现在时。惯用语，如"in/over the last decades""currently"和"in recent years"通常用来表示时间。

2. 以前的研究

第二步是在相关领域提供相关的研究工作。在介绍前人的研究著作时，作者通常用一般现在时强调当前关注的问题，用一般过去时强调其他研究者所取得的成就和局限性。

3. 综述的内容

这一步说明了这篇综述要包含的内容，因此使用现在时态和将来时态，有时也使用过去时态。

4. 意义和/或对未来研究的建议

作为一个可选的举措，作者可能会强调本综述的重要性，并对未来的研究提出建议。这里用的是现在时态和将来时态，以及虚拟语气。

（二）惯用语

综述型论文引言部分通常会使用一些惯用语，以帮助读者了解论文的目的、范围和结构。有一些惯用语可以帮助您识别不同的语步，这些惯用语包括但不限于以下内容：

1. （研究背景）的惯用语

（1） In the field of plant breeding...

(2) In recent years, there has been a growing interest in the study of...

(3) Over the past decade, there has been significant progress in understanding...

(4) In the past decades, the development of... has captured a lot of attention.

(5) One important challenge is...

2. (以前的研究) 的惯用语

(1) Previous research has shown that... however, there is a gap in our understanding regarding...

(2) Research on this topic dates back to the early 20th century when... Despite advances in research on this topic, there are still many unanswered questions about...

(3) Despite extensive research on this topic, many questions remain unanswered...

(4) This research builds upon earlier work by...

3. (综述的内容) 的惯用语

(1) This paper aims to fill this gap by...

(2) This article provides a detailed review of the research conducted on.

(3) This review provides a concise summary of recent advances in...

(4) This paper/article/review will concentrate/focus on...

4. (意义和/或对未来研究的建议) 的惯用语

(1) Understanding this topic is critical for addressing key challenges facing... today.

(2) This research is critical for developing effective solutions to pressing... problems.

(3) This paper will provide researchers with...

(4) The main contribution of this paper is...

(5) In this article/paper/review, we make readers aware of the possibilities of...

三、举例分析

(一) 举例1

Perennial cotton ratoon cultivation: A sustainable method for cotton production and breeding

(1) Cotton is an industrial textile crop and the most widely grown natural fiber crop (Zhang et al., 2022). (2) Excellent varieties are the basis for high cotton yields, especially those developed in the 1990s, when the breeding of Bt cotton saved cotton production, which 句子 (1) ~ (3) 介绍了研究背景。

was almost destroyed by bollworm. (3) This advancement also reduced the use of chemical pesticides, saving both human and material resources and contributing to environmental protection and the ecological balance, and was therefore rapidly promoted and applied in production. (4) However, Bt-transgenic insect-resistant cotton only has a particular effect on some *Lepidoptera* pests and is not very effective against aphids, sooty mites, and other pests that are currently causing severe damage in production (Bergman, 1985; Silva et al., 2008; Wan et al., 2017). (5) In addition, as labor inputs are higher for cotton than food crops, while the economic efficiency is low, and cotton is planted in drought-affected and saline areas as well as mudflats, the breeding of cotton varieties with multistress resistance has important and far-reaching significance.

句子（4）和（5）引入了存在的问题。

(6) Given the narrow genetic basis of high-yield cotton varieties worldwide and their high degree of homogeneity, as well as the controversy over the ecological safety risks associated with genetically modified cotton, it is extremely important to use semiwild lines of cultivated species and wild species, which make up the majority of *Gossypium*, as germplasm resources to expand the genetic basis of annual cotton varieties (Teravanesyan and Belova, 1970; Wang, 2007; Migicovsky and Myles, 2017). (7) In the future, it will be necessary to utilize the excellent traits of perennial species and their underlying genes.

句子（6）和（7）提出了解决问题的措施，即本文要综述的内容。

(8) The high regeneration potential of cotton has long been recognized, with examples of ratoon cotton being grown in Georgia in as early as 1786 (Seabrook, 1844). (9) In 1961, Stroman proposed ratooning F_1 plants in Peru to harvest more than one crop of F_2 seeds (Stroman, 1961). (10) In 1968, Weaver proposed that ratooning male-sterile plants would provide an excellent way to produce F_1 seeds (Weaver, 1968). (11) In the practice of cotton breeding, ratoon cotton could be used to extend the time in which germplasms can be

句子（8）~（11）介绍了以前的研究。

utilized for more than one season (Muhammad et al., 2015). (12) In tropical cotton production areas, the most economical way to take advantage of ratoon cotton is to produce hybrid cotton seeds with high quality and low cost by ratooning male sterile lines because if ratoon cotton is used for lint production in the tropics, its economic benefit is far lower than that of hybrid seed production and even worse than that of cotton sown annually in temperate areas.

句子（12）说明了研究的意义。

（二）举例2

(1) Grafting is an agricultural practice that integrates two or more plants into a single plant, and it has been applied for more than 2,000 years (Mudge et al., 2009; Zhang et al., 2020a). (2) In an individual grafted plant, the rootstock is usually in a lower position and functions as the root, whereas the scion is on the top and functions as the canopy. (3) In general, shoots and buds are used as scions in traditional grafting. (4) With the development of grafting techniques, the leaf, inflorescence, ovary, stigma, embryo, callus, and even cells can be used as scions for in vitro grafting (cell grafting or micrografting) (Zhang et al., 2018a).

句子（1）~（4）介绍了研究背景。

(5) Generally, grafting is widely used in horticultural crops such as fruit, vegetable, and ornamental (Nawaz et al., 2016); however, it is also extensively used in the study of non-horticultural crops, such as *Arabidopsis thaliana* (Melnyk et al., 2018), tobacco (Hu et al., 2019), and cotton (Ullah et al., 2017). (6) A literature search found that, Harland (1916) first reported on a grafting technique used to improve the resistance to cotton leaf-blister mite, and Weiss (1930) first proposed the use of graft hybridization to obtain graft-hybrids of cotton. (7) Different grafting methods include cleft grafting (Luo and Gould, 1999; Wang et al., 1999a; Wang et al., 1999b; Jin et al., 2006), splice grafting (Akhtar et al., 2002; Wang et al., 2007; Zhang et al., 2015a; Wang et al., 2019a), side

句子（5）~（7）介绍了以前的研究。

grafting (Luo and Gould, 1999), bark grafting (Wu et al., 2006), oblique cut grafting (Zhou, 2010) and T‑budding (Güvercin, 2017) used in cotton.

(8) As a bulk crop, cotton has a larger acreage and influence than any other horticultural crop. (9) Once grafting is widely used in cotton cultivation and breeding, it would have greater potential and value than its application in any other horticultural plant. (10) Moreover, due to the uniqueness of cotton species, cotton grafting will also develop many unique and specific methods, such as breaking the bottleneck of developing regenerated cotton seedlings from chimeras of distant grafting. (11) Nevertheless, the wide application and in‑depth study of grafting on horticultural crops have many important references for its application on cotton, such as using grafting machines to improve the efficiency of cotton grafting.

句子(8)~(10)说明了研究的意义。

句子(11)提出了对未来研究的建议。

(12) Currently, the primary purpose of cotton grafting is to improve a plant's resistance to various stresses and increase the yield of the scion (Zhang et al., 2013a; Ullah et al., 2017). (13) Grafting is also applied to preserve and propagate special germplasm resources (Hsi et al., 1955; Shen and Liu, 1979; He and Chen, 1981) and to resolve the difficulties in the heterosis utilization of cotton (Zhang et al., 2013a; Zhou, 2016). (14) In particular, annual cultivated cotton can be grafted on seedlings or ratooned crops of perennial species for achieving perennial growth in frost‑free regions. (15) Additionally, grafting may also induce inheritable variations that can be utilized to produce new germplasm resources (Stegemann and Bock, 2009; Hao et al., 2014). (16) For the molecular genetic improvement of cotton, grafting can be utilized to resolve the difficult problem of transplanting regenerated plants and transgenic plants (Wang et al., 1999a; Zhu and Sun, 2000; Akhtar et al., 2002; Chaudhary et al., 2003; Jin et al., 2006; Wang et al., 2019a). (17) Furthermore, grafting is

句子(12)~(17)本文要综述的内容。

an effective method of studying signal transduction and substance transport between roots and shoots (Kong et al., 2012; Li et al., 2012; Wang et al., 2012; Xia et al., 2013; Luo et al., 2019; Yang et al., 2021). (18) Overall, grafting has been utilized in the fields of cotton production, breeding, and basic research.

 阅读练习

阅读下面一篇评论论文的引言,并根据本单元中提到的4个语步进行分析。把句子的序号填放在表5.4右栏中。

Salinity is one of the major factors limiting agricultural production. Feeding over 9.3 billion populations in 2050 is a big challenge for agriculture. It is estimated that in 2050, agricultural production needs to be increased over 60% at 2005-2007 level (Fita et al., 2015). However, efficient agricultural production is always threatened by stress conditions such as salinity. In recent years, climate change, seawater backflow, groundwater infiltration, human-being activities such as irrigation and fertilizer application increased salt concentration in the surface soil, resulting in soil salinization (Zhang et al., 2016; Qian et al., 2021). Soil salinization inhibits plant growth, yield and product quality (Yang and Guo, 2018a). Around the world, more than 950 million hectares of land are affected by salinity stress, and soil salinization shows an increasing trend (Yang and Guo, 2018b).

The main components of salt stress in plants are osmotic stress, ionic stress, and secondary stress i.e., ROS over-accumulation (Parihar et al., 2015; Morton et al., 2018). Firstly, upon the onset of salt stress, high salinity reduces the water potential around the plant roots, limiting root absorption of water (Negrão et al., 2016). Secondly, over-accumulation of sodium and chloride in plants causes ion toxicity. It not only disrupts ion homeostasis such as Na^+ and K^+ homeostasis (Zhu, 2002), but also hinders the efficient uptake of nutrient elements such as Ca^{2+}, resulting in the lack of essential nutrients in plants (Zhang et al., 2017; Wu, 2018a; Li et al., 2021). Osmotic and ionic stresses lead to over-accumulation of reactive oxygen species (ROS) in plants causing oxidative stress (Zhu, 2016; Wu et al., 2018b). For example, excessive superoxide anion (O_2^-) and hydrogen peroxide (H_2O_2) are accumulated in chloroplasts and mitochondria, affecting photosynthesis and respiration of plants under salt stress (Balal et al., 2011). Moreover, the structure of macromolecules such as DNA and protein can be damaged by excessive ROS (Hu et al., 2021; Liu et al., 2021a). Nowadays, besides genetic engineering and exogenous application of antioxidants, nanomaterials showed its potential in improving plant salt tolerance although the underlying mechanisms are less addressed. Nano-enabled plant salt tolerance could be an alternative approach to help to enable efficient agricultural production.

In this review, we summarized the known molecular mechanism underlying plant salt tolerance and emphasized the importance of nanotechnology in improving plant salt tolerance. We hope this review will set up an idea to help the researchers in plant science and nanoscience to better understand possible new approaches to address issues such as salinity in agriculture.

表 5.4　综述型论文前言阅读方法练习

综述前言的 4 个语步	对应的句子序号
（1）语步 1——研究背景	
（2）语步 2——以前的研究	
（3）语步 3——综述的内容	
（4）语步 4——意义和/或对未来研究的建议	

第五节　主体

一、概述

综述型论文的主体部分是论文的重点。一篇综述型论文的主体部分通常包括以下内容：

（一）文献综述

系统性地回顾和总结相关文献中的观点、理论、证据或方法，以支持或反驳本研究讨论的问题或话题。

（二）分析和评价

对所涉及文献进行分析和评价，以揭示其优点、局限性、认知偏差或缺陷等方面，同时也考虑文献间的关系和一致性。

（三）综合总结

将文献综述、分析和评价的结果进行综合，得出本研究的主要观点、结论或建议。需要注意的是，不同类型的综述型论文可能会在主体部分中强调不同的方面。例如，系统综述的主体部分可能更侧重于对先前研究的方法和结果进行系统和客观的回顾和总结，而理论综述的主体部分则可能更侧重于对当前领域的理论框架进行深入的分析和评估。此外，系统综述在主体部分，还会详细说明研究方法和论证流程等方面内容。

这3个部分通常使用不同的写作方式：

（一）时间顺序的方式

作者按照时间顺序来说明围绕某一研究课题的演变过程，以及不同阶段的研究水平和成果、存在的问题和发展趋势，使读者能够了解该学科的动态发展。

（二）水平的方式

作者按照水平顺序，通过描述已解决和未解决的问题、可能的解决方案、存在的争论、各种观点，以及自己的立场、成就、发现等，对该领域的研究现状进行分析。说明这些的主要目的是指出未来的研究需求。

（三）时间与横向相结合

作者提供了某一学科发展的时间顺序，并对该学科在特定阶段的特点、发现和发展趋势等相关信息进行了横向说明。目的是强调存在的问题，并提出今后的研究方向。

二、阅读技巧

在阅读一篇综述型论文的主体部分时，请遵循每个小节的小标题。小标题为理解综述的内容提供了指导。通过阅读小标题，您可以了解所综述的内容。下面是一个例子。

示例：

1　Use of ratoon cotton for the production of commercial F_1 hybrid cotton seeds

1.1 Use of ratoon cotton in the production of F_1 hybrid cotton seeds by hand-emasculation

1.2 Ratooning male-sterile cotton for the production of F_1 hybrid cotton seeds

1.2.1 Maintenance of male sterility of sown cotton by ratooning

1.2.2 Maintenance of the male sterility of cutting-propagated plants by ratooning

1.2.3 Maintenance of male sterility of grafted plants by ratooning

三、举例分析

1. Perennial conservation of *Gossypium* species and their classification

Of the *Gossypium* species, four are cultivated species, and the others are perennial wild species (图 5.1). The basis of cotton breeding is the collection and preservation of genetic germplasms, which provide an essential foundation to improve and sustain cotton production (Zhang et al., 2012a). Specifically, the following extraordinary accessions are required for perennial preservation: (1) wild cotton species with short-day flowering behavior, from which it is hard to obtain seeds (Percy et al., 2014); (2) nullisomic, monosomic, telomeric, trisomic, and translocation lines and other cytogenetic stocks of cotton (Kiranga, 2013), the fruiting rate of which is low and the identification of which is difficult and time-consuming; (3) cotton plants infected by some kinds of pathogens, the status of which should be maintained for a long time for research (Mihail et al., 1987; Seo et al., 2006); and (4) cotton hybrids or backcross generations such as F_1, F_2, and BC_1F_1 generations, which can be commonly used for only a year but the generations of which can be repeatedly used over many years through perennial growth (D'Eeckenbrugge and Lacape, 2014).

1.1 Global overview of perennial germplasms and perennial conservation of cotton

Worldwide, *Gossypium* germplasms with different ecological niches have much morphological, agronomic, physiological, and genetic variability that is conserved *in situ* at centers of cotton origin (Castro et al., 2016) and preserved *ex situ* with a large number of accessions in eight major countries with extensive cotton germplasm collections, namely Australia, Brazil, China, France, India, Russia, the USA, and Uzbekistan (Abdurakhmonov, 2007; Rahmat et al., 2014; Boopathi and Hoffmann, 2016), attaching great importance to the preservation of perennial germplasms. It is worth noting that more than 20,000 cotton germplasm accessions are preserved in Uzbekistan, making that the most extensive such collection in the world. Although cotton is now rarely

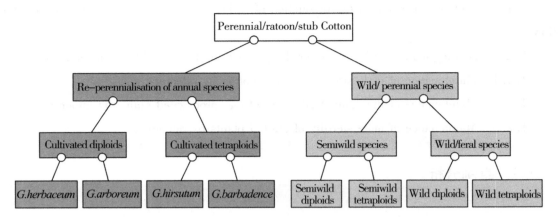

图 5.1 Classification of ratoon cotton based on cultivars, semiwild species, wild/feral species, and chromosome ploidy. The dark gray boxes show the cultivated species, and the light gray boxes show the wild species.

grown in France, it is commendable that more than 3,000 accessions, including approximately 1,000 wild accessions, are preserved in CIRAD (Coopération Internationale en Recherche Agronomique pour le Développement), a French publicly supported agency that specializes in tropical and Mediterranean agriculture (Campbell et al., 2010).

At present, China, India, and the USA are the three largest cotton-producing countries. Of the approximately 10,000 accessions preserved in the National Cotton Germplasm Collection (NCGC) of the USA, 581 are wild germplasms. Most of the accessions, including photoperiodic germplasms and perennial accessions, are perennially grown at the tropical Cotton Winter Nursery (CWN) in Tecoman, Colima, Mexico (Wallace et al., 2009; Percy et al., 2014). In India, the bank of the Central Institute for Cotton Research has collected a total of 10,227 accessions, including 26 wild species and 32 perennial forms (Boopathi et al., 2014). Although China is not an origin center of cotton, most of the 8,868 accessions, including 32 wild forms, were collected from China, and 2,236 accessions were introduced from 52 foreign countries. The Sanya National Research Station of Wild Cotton Germplasm, which is located within the tropics of China, has 391 wild accessions that are perennially grown for conservation (Jia et al., 2014). In addition, Mexico, Pakistan, and other cotton-planting countries have also collected many germplasms.

1.2 Gene pools and genome assignment of perennial *Gossypium* species

At least 48 species of *Gossypium*, including 7 tetraploid ($2n = 4x = 52$) species, 41-45 diploid ($2n = 2x = 26$) species, and other wild species, originate from arid to semiarid regions within the tropics and subtropics (Wendel and Grover, 2015; Shim et al., 2018; Wang et al., 2018). According to the genetic relationship with upland cotton, all cotton species can be classified into primary, secondary, and tertiary gene pools. Of the 7 tetraploid species (genome AD), *G. hirsutum* and *G. barbadense* are mainly cultivated worldwide, so all tetraploid species are in the primary gene pool. Based on the relative genetic approachability and utility of species to improve *G. hirsutum* and *G. barbadense*, 20-21 diploid species (genome A/D/F/B) and 21-24 diploid species (genome C/E/G/K) have been classified into secondary and tertiary gene pools, respectively (图 5.2).

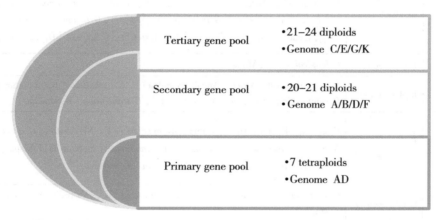

图 5.2 The primary, secondary, and tertiary gene pools based on their genetic relationship with upland cotton. The primary, secondary, and tertiary cotton (*Gossypium*) gene pools are shown from the inside circle to the outer ring. The farther away the primary gene pool is, the further the genetic approachability is from the tetraploids, and the richer the genetic diversity.

Gossypium species with genomic assignments and geographical origins in the primary, secondary, and tertiary cotton gene pools are detailed in 表 5.5. Responding to the diverse geographic and ecological conditions of frost-free regions, wild cotton species show a broad adaptation range from herbaceous perennial diploid species with a fire-/dry-adapted biseasonal growth pattern in northwest Australia to small cotton trees dropping their leaves to avoid the effects of the dry season in southwest Mexico (Campbell et al., 2010). Therefore, it is widely believed that the extensive genetic diversity within wild cotton increases their opportunities for evolutionary adaptation that reduces their genetic vulnerability to changing harmful environments (Boopathi and Hoffmann, 2016).

表 5.5 *Gossypium* species with genome assignment and geographic origin in the primary, secondary, and tertiary gene pools.

Gene pool	Genome	Presently recognized species in *Gossypium* [Genome assignment, geographic origin]
Primary (7 tetraploids)	AD (7 species)	*G. hirsutum* [(AD)$_1$, Central America], *G. barbadense* [(AD)$_2$, South America], *G. tomentosum* [(AD)$_3$, Hawaiian Islands], *G. mustelinum* [(AD)$_4$, Brazil], *G. darwinii* [(AD)$_5$, Galapagos Islands], *G. ekmanianum* [(AD)$_6$, Dominican Republic], *G. stephensii* [(AD)$_7$, Wake Atoll]
Secondary (20–21 diploids)	A (2 species)	*G. herbaceum* [A$_1$, Southern Africa] (subs. *africanum* [A$_{1-a}$, Southern Africa]), *G. arboretum* (syn. *G. aboreum*) [A$_2$, Indus valley, Madagascar]
	D (13–14 species)	*G. thurberi* [D$_1$, Mexico], *G. armourianum* [D$_{2-1}$, Mexico], *G. harknessii* [D$_{2-2}$, Mexico], *G. davidsonii* [D$_{3-d}$, Mexico], *G. klotzschianum* [D$_{3-k}$, Galapagos Islands], *G. aridum* [D$_4$, Mexico], *G. raimondii* [D$_5$, Peru], *G. gossypioides* [D$_6$, Mexico], *G. lobatum* [D$_7$, Mexico], *G. trilobum* [D$_8$, Mexico], *G. laxum* [D$_9$, Mexico], *G. turneri* [D$_{10}$, Mexico], *G. schwendimanii* [D$_{11}$, Mexico], {*G.* sp. nov. [D$_{12}$, Mexico]}[a]
	F (1 species)	*G. longicalyx* [F$_1$, Africa]
	B (4 species)	*G. anomalum* [B$_1$, Africa (Angola, Namibia)], *G. triphyllum* [B$_2$, Namibia in Africa], *G. capitis-viridis* [B$_3$, Cape Verde Islands], *G. trifurcatum* [B, Somalia]
Tertiary (21–24 diploids)	E (4–7 species)	*G. stocksii* [E$_1$, East Africa, Arabia], *G. somalense* [E$_2$, NE Africa], *G. areysianum* [E$_3$, Arabia], *G. incanum* [E$_4$, Arabia], {*G. benadirense* [E, Somalia, Ethiopia, Kenya]}[b], {*G. bricchettii* [E, Somalia]}[c], {*G. vollesenii* [E, Somalia]}[d]
	C (2 species)	*G. sturtianum* [C$_1$, Australia] (var. *nandewarense* [C$_{1N}$, Australia]), *G. robinsonii* [C$_2$, Australia]
	G (3 species)	*G. bickii* [G$_1$, Australia], *G. australe* [G$_2$, Australia], *G. nelsonii* [G$_3$, Australia]
	K (12 species)	*G. exiguum* [K$_1$, Australia], *G. rotundifolium* [K$_2$, Australia], *G. populifolium* [K$_3$, WA Australia], *G. pilosum* [K$_4$, WA Australia], *G. marchantii* [K$_5$, Australia], *G. londonderriense* [K$_6$, Australia], *G. enthyle* [K$_7$, Australia], *G. costulatum* [K$_8$, Australia], *G. cunninghamii* [K$_9$, Northern NT Australia], *G. pulchellum* [K$_{10}$, WA Australia], *G. nobile* [K$_{11}$, Australia], *G. anapoides* [K$_{12}$, Australia]

Note: The data came from Shim et al. (2018), Wang et al. (2018), and Wendel and Grover (2015). The farther the genome from the primary gene pool is, the further the genetic approachability. [a] No formal name yet, [b,c,d] no living materials of these species have been available for further study (Wang et al., 2018).

1.3 Efficient strategies of collecting, conserving and characterizing perennial cotton germplasm

Perennial cotton germplasms can be widely collected through exploration in tropical origin centers or via exchange with other gene banks. Perennial cotton can be conserved *in situ* or *in vi-*

vo in tropical fields. After harvesting enough seeds, the seeds can be preserved in the gene bank. Moreover, perennial cotton germplasm can be preserved as perennial roots and cuttings and via grafting and tissue culture in greenhouses or laboratories. Most traits can be evaluated in tropical areas, and some abiotic or biotic stress responses can be characterized in greenhouses. Molecular biological methods, such as genomics, phenomics, and molecular markers, can be used to improve the efficiency of characterizing perennial cotton germplasms.

分析：

通过阅读小标题，您可以了解到，在本节中，作者希望介绍棉属种质的多年生保护及其分类，先对再生棉进行了分类，然后总结了哪些棉花种质需要多年生保护，继而通过3个副标题详细描述了这两方面内容。在2.1下，对棉花多年生种质资源和多年生保护的全球概况进行了介绍；在2.2下，对多年生棉属植物的基因池和基因组分配进行了详细说明；在2.3下，综述了收集、保存和鉴定多年生棉花种质的有效策略。

同样的组织方式也适用于其他部分。

 阅读练习

阅读以下综述型论文的主体，并尝试根据小标题总结所查看的内容。

2. Grafting used for cotton production

In cotton production, an adequate grafting rootstock can change the original relationships between the root and canopy, the resistance of the scion (Akhtar et al., 2013; Zhang et al., 2013a; Ullah et al., 2014), and the utilization efficiency of mineral nutrients and water (Kong et al., 2012; Li et al., 2012; Wang et al., 2012; Xia et al., 2013; Kong et al., 2017; Luo et al., 2019), resulting in improved yield. Annual cotton cultivars can be grafted onto wild cotton for perennial cultivation in frost-free regions where survival in open fields is difficult (Zhang et al., 2013a; Zhou, 2016). In general, compared with upland cotton, sea island cotton presents stronger resistance, more vigorous growth, and superior fiber quality but a lower yield and more extended growth period; therefore, sea island cotton is often used as the rootstock and upland cotton as the scion. Bt (*Bacillus thuringiensis*) transgenic cotton has been used as rootstock because it has stronger resistance to cotton bollworms than non-transgenic cotton (Rui et al., 2005); furthermore, the Bt protein content in the roots of transgenic hybrid cotton was much higher than in the stems (Yu et al., 2015). Examples of resistant stocks used in grafted cotton for different stresses are shown in 表 5.6.

表 5.6 Resistant stocks used for different stresses in grafted cotton.

Stress	Class	Component	Resistant stocks	References
Biotic	Fungi	*Verticillium* wilt	*G. barbadense* 'Hai 7124 & Pima 90'	Hao et al. (2010); Zhang et al. (2012a)
			G. arboreum Shixiya	Lou (2010)
			G. barbadense 'Hai 7024'	Zhang et al. (2018b)
	Virus	Leaf curl disease	*G. arboreum* Ravi	Akhtar et al. (2013); Ullah et al. (2014)
			G. hirsutum AS0039 and AS0099	Ullah et al. (2017)
			G. herbaceum 'Co Tiep Khac'	Ullah et al. (2018)
	Pest	Cotton bollworm	*G. hirsutum* 'SGK321 & 99B'	Rui et al. (2005)
			G. hirsutum 'CCRI41'	Jin (2013)
		Leaf-blister mite	*G. hirsutum* in Barbados	Harland (1916)
		Root-knot nematode	*G. hirsutum* Auburn 56	McClure et al. (1974)
Abiotic	Salt	Low potassium	*G. hirsutum* 'SCRC22'	Li et al. (2012); Wang et al. (2019b)
			G. hirsutum 103	Xia et al. (2013)
			G. hirsutum 'CCRI49'	Zhang et al. (2017)
		NaCl	*G. hirsutum* 'SCRC28'	Kong et al. (2012); Xia et al. (2013)
			Perennial *G. barbadense* 113-5	Qiu et al. (2015)
	Water	Drought	*G. hirsutum* 'K836'	Luo et al. (2019)
	Temperature	Chilling	Perennial *G. barbadense*	Zhang et al. (2013a); Zhou (2016)

...

2.1 Improving scion resistance to biotic stresses

Grafting is a suitable method for improving cotton resistance to disease and pests, such as *Verticillium* wilt, cotton leaf curl disease (CLCuD), cotton bollworm, and root-knot nematode (McClure et al., 1974).

...

2.2 Enhancing tolerance of grafted cotton to abiotic stresses

Plants uptake mineral nutrients and water for shoot growth mainly via the root, and grafting can change the original relationships of the root-canopy and subsequent hormonal signals and substance transport in the grafted plant, which results in changes in leaf senescence (Kong et al., 2012; Li et al., 2012; Wang et al., 2012). For example, with or without potassium (K) application, the proportion of dry matter and K in vegetative organs decreased, dry weight and K in reproductive organs increased, and yield and K use index increased in the self-grafted inefficient genotype 122 compared with ungrafted seedlings (Xia et al., 2012).

…

2.3 Grafting annual cotton for perennial cultivation in frost-free regions

All kinds of wild cottons are perennial species and have stronger resistance to different stresses. Moreover, they show a greater over-wintering survival rate than annuals, although perennials have a lower yield and a more extended growth period over annuals. Annual cultivars with high yield can be ratooned into perennial plants in the frost-free regions where the average temperature is above 10℃, and the minimum temperature is above 0℃ in the coldest month. In 2011, an application for a US patent was made for a method of grafting specific annual cultivars that cannot survive over winter onto perennial rootstocks to achieve perennial growth, and it was granted on 2016-02-23 (Zhou, 2016). Subsequently, perennial cultivation areas of cotton have been extended from tropical region to near tropics by this method, which can also be used for heterosis fixation. Moreover, the yield of grafted hybrid F_1 cotton can be significantly increased because of the lengthened growth period of the scion (Zhang et al., 2020b).

第六节　结　论

一、概述

综述型论文的结论部分通常是对所综述内容的总结，旨在表达作者的观点和思考，对综述的学术领域进行评价，指出该领域存在的挑战，并展望未来的研究前景。

一篇综述型论文的结论通常包括以下 5 个语步：

（一）总结

对论文中讨论的主要观点和结论进行总结。

（二）评价

强调论文中讨论的话题的重要性，并概括出他们对相关领域的意义。

（三）限制

指出论文在研究或分析过程中所遇到的任何限制或局限性。

（四）建议

在合适的情况下，提供实际建议以帮助读者更好地理解研究领域或如何开展后续研究。

（五）展望

提供未来工作或研究的展望，可能包括进一步研究或相关问题的探讨。

需要注意的是，不同类型的综述型论文可能会在结论部分中强调不同的方面。例如，系统综述的结论可能会更侧重于总结和评价先前研究的证据，而理论综述的结论则可能更侧重于对当前领域的理论框架进行总结和建议。

二、阅读技巧

综述型论文结论中的惯用语可能包括 review、summarize、in conclusion、overall、to sum up、ultimately、lastly、in summary、all things considered 以及 finally 等短语。这些短语用于向读者发出信号，表明作者正在总结综述的要点，并结束论文。评价性语言的使用，如 important、significant、crucial，也可以作为综述型论文结论中的惯用语，这些短语通常表示评论人正在总结他们对综述内容的总体印象。对于综述对象的限制性因素，作者往往使用 limitation、challenge、barrier、constraint、only 等词汇来表达。此外，为了指出综述内容的未来研究方向，综述结论末尾的句子中往往出现 future、further、deeper 等惯用语。

在阅读一篇综述型论文的结论部分时，对于上面概述中提到的 4 个语步，可以通过每个语步的一些惯用语找出相应的内容。下面分步举例说明。

（一）总结的惯用语

（1）This review provided/offered a detailed summary of the research conducted on...

（2）This review summarized/concentrated on/focused on...

（3）We have summarized/reviewed the recent development of...

（4）We summarized/reviewed the current state of...

（5）A systematic/thorough review has been performed following...

（6）The findings lend/offer support to...

（二）评价的惯用语

(1) We believe/consider/feel/hypothesize/think that...
(2) It is believed/proposed/recommended that...
(3) The author's/Our opinion is that...
(4) It is suspected/doubted that...
(5) It is important/believable that...
(6) It seems that... can account for/interpret this...

（三）限制的惯用语

(1) The challenge is that...
(2) This review has only addressed the issue of...
(3) The problem remains as to how...
(4) To overcome...

（四）建议的惯用语

(1) ...need to/should be further explored/investigated.
(2) It is necessary to make/carry out/conduct/develop...
(3) Deeper/more/further study on... needs to be done.
(4) Follow-up study/research will further clarify/confirm...

（五）展望的惯用语

(1) In the future...
(2) It can be expected/believed that...
(3) Follow-up study/research will further clarify/confirm...

三、举例分析

（一）例1

(1) The suitable rootstocks not only enhance soil resource utilization but can enhance resistance to biotic and abiotic stresses leading

句子（1）和（2）重述了综述的内容（特

towards improved cotton production. (2) It is also important to understand the mechanisms of root-shoot communication, gene silence, transferring, and junction formation (Figure. 5.3). (3) In the future, when grafting in cotton is widely adopted in the production like a horticultural plant, simple, efficient, and mechanized grafting technique needs to be developed. (4) For distant grafting breeding in creating new and excellent germplasm, not only junction chimera, but also inheritable epigenetics of scions need to be further explored.

别是图例），并显示了作者对综述内容的观点。

句子（3）表明了作者的展望。句子（4）提出了建议。

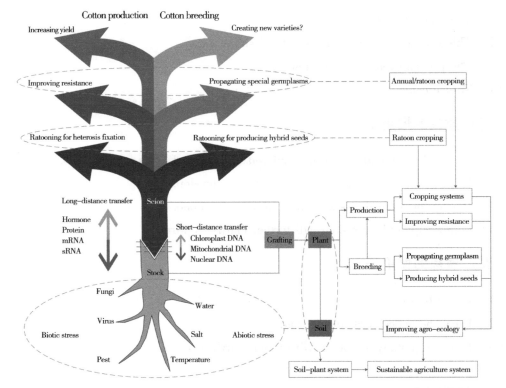

图 5.3　Substances exchange between rootstock and scion and its application in cotton production and breeding. Rootstocks that are resistant to cotton pathogens (especially fungi and viruses), pests, temperature, salinity, and water stress can increase scion resistance through the long-distance transmission of water, nutrient elements, and plant hormones, thus increasing yield in cotton production. Grafting can be used in cotton breeding to propagate special germplasms, fix heterosis, and reduce the cost of producing hybrid cotton seeds. In addition, the long-distance transfer (such as mRNA and sRNA) and short-distance transfer (chloroplast DNA, mitochondrial DNA, and nuclear DNA) of genetic material between rootstock and scion provide the possibility for genetic regulation of cotton growth and development and breeding of new cotton varieties. https://doi.org/10.1016/j.indcrop.2021.114227

（二）例 2

（1）Ratoon cropping is highly important to cotton production, the permanent maintenance of the male-sterile line for heterosis utilization, the fixation of heterosis and the preservation and generation of novel germplasm. （2）Therefore, taking advantage of the warm climate to exploit the perennial and indeterminant growth habits of cotton would be profitable, resource-saving and environmentally friendly in the tropics. （3）However, increased investments are needed for ratoon cotton breeding, cropping and agro-ecological research.

（4）Although ratooning of annual cotton for heterosis breeding has been successful in some countries, such as Australia, China and India, owing to the lack of results from evaluations of commercial production and sales of hybrid cotton seeds from ratooning systems, we can only hypothesize that this practice has good future prospects. （5）In addition, the expansion of ratoon cotton would not necessitate the use of more natural and seminatural land for agricultural development because if tropical ratoon cotton is used for lint production, its economic benefit is much lower than that of hybrid seed production or even lower than that of temperate annual cotton, and the management of ratoon cotton used to produce hybrid seeds necessitates more labor than that needed for grain crops.

句子（1）重述了综述的内容。句子（2）显示了作者对综述内容的评价。

句子（3）指出了未来的研究方向。

句子（4）指出了当前研判的限制因素。

句子（5）强调了所综述对象的主要优点和建议。

 阅读练习

阅读以下综述型论文的结论，并根据本单元中提到的 5 个语步进行分析。把句子的序号放在表 5.7 或表 5.8 右栏。

结论 A

（1）Plants mainly reduce oxidative stress induced by salt stress by reducing ROS content and increasing antioxidant activity, maintaining the ionic balance through Na^+ sequestration or extrusion, and interactions such as phytohormones and signaling transduction ultimately response to salt stress. (2) Nanomaterials play a positive role in plant salt tolerance, and the relevant mechanism needs to be further explored. (3) To date, nano-enabled plant salt tolerance is still largely in the laboratory research stage. (4) To facilitate the adoption of nano-enabled plant salt tolerance in agricultural production, discussions and set up of widely accepted policies and regulations are urgently called on task. (5) Also, it is worthy to be investigate how to accurately and economically apply NMs to a large number of plants. Leaf spraying seems to be an easy way to complete the existing infrastructure, and can really be large-scale targeted application of NMs, whereas the impact of the weather and plant leaves on the absorption efficiency of NMs must be considered. (6) Therefore, it is necessary to design NMs for agriculture in a targeted way, while adjusting the applied NMs system such as adding surfactants. (7) In addition, it is necessary to carefully consider the type of nanomaterials and fully understand not only the toxic effects of NMs on plants but also whether it is harmful to the environment and whether it can be recycled and many other factors. (8) It is worth noting that, the optimal concentration and application effect of the same NMs for different species are different, so more research needs to be considered before large-scale application. (9) We believe that nanotechnology can and will play an important role in the future of agriculture, including maximizing the resilience of crops to increase global food production.

表 5.7 综述型论文结论阅读方法练习 1

综述结论的 5 个语步	对应的句子序号
(1) 语步 1——总结	
(2) 语步 2——评价	
(3) 语步 3——限制	
(4) 语步 4——建议	
(5) 语步 5——展望	

结论 B

(1) This review of ratoon cotton identifies immediate and long-term research requirements. (2) The challenges of increased disease and pest incidence in ratoon cotton need to be addressed with urgency. (3) We were not able to identify whether *H. armigera* contributed to

the abandonment of arboreal cotton ratoon cultivation in the mid-20th century in southern China. (4) Some studies from tropical India that we reviewed suggest that Bt cotton, which is resistant to insect pests, shows higher production than non-Bt cotton. (5) More studies are necessary to evaluate the net benefits between ratoon and sown cotton cropping to provide flexible risk management strategies. (6) Such actions may stimulate greater support for the sustainable and innovative development of ratoon cotton cultivation. (7) During periods of persistent labor shortages in most villages, ratoon cultivation offers farmers a way to increase their income with minimal added effort. (8) On-farm studies are particularly important to determine the optimum planting density under local conditions. (9) It is worth mentioning that ratooning is particularly compelling in arid and semi-arid ecosystems as a soil and water management strategy. (10) Moreover, ratoon cotton may be a good option for rehabilitating land contaminated by pesticides or heavy metals that are not suitable for planting food crops.

(11) Base on the three methods for ratooning cotton that were previously discussed, perennial ratoon cropping of semi-wild cotton has the best ecological benefits, perennial ratoon cropping of annual cultivars has comprehensive eco-economic benefits, whereas biannual ratoon cropping of annual cultivars presents the highest economic benefits. (12) Ratooning sown cotton for the second fruiting cycle can provide an additional yield of 60%~70% in the same season, avoid closed-season legislation, and has great potential and prospects for cotton production. (13) However, there has been a sharp global decline in rural populations and urbanization development has created areas with idle arable land. (14) Moreover, even if ratoon cotton was to expand, it would not develop natural and semi-natural environments into arable land since cotton is a labor-intensive crop and mechanized harvest is still difficult. (15) Consequently, these methods require multi-year testing in large-scale agricultural settings, and linking farmers with emerging markets is the next step. (16) Therefore, related services should be provided to meet farmers' interest in ratoon cultivation of cotton through participatory field trials and demonstrations. (17) Finally, studies can help to overcome the constraints that companies and farmers face with respect to the broader cultivation of ratoon cotton and improve our understanding of its economic and ecological significance.

表5.8 综述型论文结论阅读方法练习2

综述结论的5个语步	对应的句子序号
（1）语步1——总结	

（续表）

综述结论的 5 个语步	对应的句子序号
（2）语步 2——评价	
（3）语步 3——限制	
（4）语步 4——建议	
（5）语步 5——展望	

第六章

专业英文文献的阅读和翻译方法

第一节　专业英文文献的阅读方法

一、概述

文献阅读是学术研究中至关重要的一环。文献阅读可以帮助研究人员了解其领域内的现有研究,从而指导其研究方向、方法和写作。

二、专业英文文献的特点

专业英文文献的特点通常包括以下几个方面:

(一) 专业性

专业英文文献涵盖了广泛的领域和话题,因此需要具有一定的专业知识才能理解其中的内容。

(二) 精确性

专业英文文献的语言非常精确,使用术语和符号来传达特定的含义时,需要注意这些符号的定义和用法。

（三）结构化

专业英文文献通常遵循一定的结构规范，如摘要、引言、方法、结果和讨论等部分，这有助于读者更快速地找到所需信息。

（四）可重复性

专业英文文献的试验方法和数据通常都应该是可重复的，这也是科学研究的基本原则之一。

三、阅读专业英文文献的方法

在此，我们介绍几种常用的专业英文文献阅读方法，以帮助研究人员更好地利用文献阅读来指导其研究。

（一）浏览式阅读法

浏览式阅读法是指对一篇专业英文文献进行快速浏览，以了解其主题和大意，并找到其中的重点和关键信息。这种方法适用于初次接触一个领域或主题时，以及在短时间内了解一个领域内研究进展时。在浏览式阅读中，研究人员可以查看文献的标题、摘要、引言和结论等部分，以便快速了解其内容。在使用这种方法时，研究人员需要对阅读文献的目的和问题有一个明确的理解。以下是浏览式阅读的几个步骤：

（1）浏览标题和摘要：先仔细浏览论文的标题和摘要，了解论文的主题和目的，确定该文献是否与自己的研究领域相关。

（2）翻阅内容页：翻阅文献的目录和章节标题，了解论文的结构和组织方式，找到自己感兴趣的部分。

（3）快速浏览正文：快速浏览正文，留意论文中的关键词、段落标题和粗体字等格式特征，找到论文的重点和关键信息。

（4）查看表格和图片：查看论文中的表格和图片，了解试验结果和数据变化趋势，找到论文的重点和关键信息。

（5）总结主要内容：根据浏览的结果，总结出论文的主要内容和观点，帮助理解论文的核心思想和论证过程。

通过浏览式阅读法，能够在短时间内了解该领域的研究进展和问题，发现有价值的信息和资源，为自己的研究工作提供参考和指导。同时，也能够帮助新手掌握该领域的基本知识和概念，提高阅读速度和阅读效率。需要注意的是，浏览式阅读法不能替代详读式阅读法，对于重要的文献还需要进行仔细地阅读和分析。

（二）扫描式阅读法

扫描式阅读法是指快速扫视一篇专业英文文献，对文献进行有选择性的阅读，了解其主题和大意，并在此基础上进行筛选和过滤，以确定是否需要深入阅读。这种方法适用于研究人员已经有了一定的领域知识，并需要进一步了解某些特定的研究问题或主题时。在使用扫描式阅读法时，研究人员可以更加仔细地阅读文献的引言、方法、结果和讨论等部分，以便获取更多的信息。此外，扫描式阅读法还可以通过阅读文献中的参考文献来进一步了解该领域内的其他研究。以下是扫描式阅读的几个步骤：

（1）浏览标题和摘要：先浏览论文的标题和摘要，了解论文的主题和目的，快速掌握论文的内容和结论。

（2）检查图表：扫描论文中的图表，了解试验结果和数据趋势，找到论文的重点和关键信息。

（3）读结论段：跳过正文的细节内容，直接阅读结论部分，了解作者的总体研究发现和观点，判断论文是否值得进一步阅读。

（4）筛选参考文献：检查论文中引用的参考文献，了解该领域的前沿研究和相关工作，筛选出有价值的文献以供进一步阅读。

（5）总结主要观点：根据扫描和筛选的结果，总结出论文的主要观点和结论，对论文的价值和适用性形成初步判断。

通过扫描式阅读法，能够帮助科研工作者和学生快速了解该领域的研究进展和问题，快速筛选出有价值的文献进行深入阅读，提高阅读效率和节约时间。

（三）详读式阅读法

详读式阅读法是指对一篇专业英文文献进行仔细、详尽的阅读，全面理解作者的观点、试验设计、方法和结论等方面的内容，并在此基础上形成自己的认识。这种方法适用于研究人员需要深入了解某个领域或主题时。在使用详读式阅读法时，研究人员需要对文献进行全面的阅读，并对其内容进行逐字逐句的分析和理解。此外，研究人员还可以将文献中的数据和结果与其自己的研究进行比较和分析。以下是详读式阅读的几个步骤：

（1）粗略浏览全文：先快速阅读全文，了解论文的主题、目的和框架，掌握整篇论文的大致思路和结构。

（2）逐段详细阅读：从论文的开头开始，逐段进行仔细阅读，深入理解作者所表达的观点和试验过程，同时注意标注重要信息和关键词。

（3）总结段落大意：阅读每一段后，总结该段的主旨和内容，整合不同段落之间的联系和转换，帮助理解整篇论文的逻辑结构和脉络。

（4）分析数据和试验：对于论文中的数据、图表和试验过程进行分析，考察试验设计的合理性、数据的可靠性和推论是否严密等方面的问题，以判断试验结果的有效性和普遍性。

（5）思考自己观点：根据阅读和分析的结果，思考自己对于论文的认识和态度，包括对某些观点的认同或异议，以及对研究方向和方法的看法。

通过详读式阅读法，能够帮助科研工作者更深入地了解该领域的研究进展和问题，提高其思考和判断能力，从而更好地指导自己的研究工作。同时也能够帮助学生更好地掌握专业知识，提高文献阅读和理解能力。

（四）批判式阅读法

批判式阅读法是指对于一篇专业英文文献进行深入思考和分析，发现其中的优点和缺点、逻辑结构和论据支持等方面的问题，并在此基础上形成自己的看法和观点。以下是批判式阅读的几个步骤：

（1）粗略浏览全文：阅读全文前先浏览标题、摘要、引言、结论和参考文献等部分，对论文的主旨和框架有一个初步了解。

（2）详细阅读全文：在第一步基础上，对文献进行仔细阅读，理解作者的研究方法、试验设计和结果等内容，并注意批判性地思考其中的问题和不足之处。

（3）分析论证结构：对文献中的论证结构进行分析，包括数据来源和分析、试验设计和方法是否合理等方面的问题，检验其逻辑严谨性并寻找可能存在的问题或漏洞。

（4）思考作者观点：针对作者的观点和论点，进行思考和比较，与已有研究和个人经验相互印证，形成自己的看法和观点，同时也应注意客观评价和避免主观臆断。

（5）总结论文价值：根据以上阅读和分析，总结出该文献的优点和不足之处、新颖性和前瞻性，并整理出个人的批判性评价和对未来方向的展望。

通过批判式阅读法，能够帮助科研工作者更全面地了解该领域的研究进展和问题，提高其思考和判断能力，从而更好地指导自己的研究工作。

四、不同研究阶段的专业英文文献阅读方法

不同研究阶段的专业英文文献阅读方法略有不同，下面简要介绍：

（一）初步探索阶段

在此阶段，需要进行广泛的文献调研，以了解该领域的研究进展和问题，确定自己的研究方向和课题。此时，可以使用扫描式阅读法，快速浏览大量文献，筛选出与自己研究相关的文献，并建立自己的文献库。

（二）深入理解阶段

在此阶段，需要深入理解所选文献的内容和结论，将其整合到自己的研究框架中，帮助设计试验或撰写论文。此时，可以使用详读式阅读法，对所选文献进行深入阅读和分析，了解作者的思路和方法，并进行思考和批判性评价。

（三）试验设计阶段

在此阶段，需要选择合适的试验方法和技术，设计试验方案，以验证自己的研究假设。此时，可以使用浏览式阅读法，查阅与自己试验方法和技术相关的文献，找到操作指南和经验总结，帮助设计试验和解决试验中遇到的问题。

（四）论文写作阶段

在此阶段，需要撰写论文，将研究结果和结论整理成系统性的文稿。此时，可以使用扫描式阅读法，查阅与自己论文主题相关的文献，找到引言、绪论、讨论等部分的参考文献，帮助论据论证和文字表达。

总之，在不同的研究阶段，需要采用不同的专业英文文献阅读方法，以满足研究需求和目标。

五、阅读专业英文文献的辅助方法

（一）记笔记

在阅读过程中，可以适当记录笔记，包括论文的主要观点、重点内容和自己的理解

等，帮助加深对文献的理解并方便后期查阅和总结。

（二）多角度思考问题

在阅读专业英文文献时，应该从多个角度思考问题，将其与其他文献和自己的实际情况联系起来，有助于更好地理解和应用。

（三）与他人讨论

与他人讨论论文也是一个非常有效的阅读方法。讨论可以帮助更好地理解论文并发现其中存在的问题，同时也可以促进自己的学习和成长。

六、阅读专业英文文献的注意事项

针对专业英文文献的特点，以下是一些阅读专业英文文献的注意事项：

（一）了解论文类型

不同类型的专业英文文献（如综述、研究型论文、案例报告等）具有不同的结构和特点，需要根据不同类型的文献采用不同的阅读策略。

（二）阅读摘要

阅读摘要可以帮助您了解论文的主题、目的、方法、结果和结论，从而判断是否值得进一步阅读全文。

（三）注意图表

专业英文文献中的图表通常非常重要，它们可以帮助您更好地理解试验方法和数据结果。

（四）着重阅读方法和结果部分

科学研究中的方法和结果部分是最关键的部分，需要认真阅读并理解其中的细节和重点。

七、提高专业英文文献阅读能力的方法

以下是提高专业英文文献阅读能力的几个方法：

（一）扩充词汇量

阅读英文文献需要掌握大量的专业术语和词汇，因此扩充词汇量是提高阅读能力的重要基础。可以通过背单词、多读英文材料等方式来增加词汇量。

（二）学习句法结构

英文句子结构较为灵活复杂，而且通常存在省略和倒装现象。因此，学习英文句法结构非常重要，这有助于理解和分析长难句。

（三）熟悉专业知识

阅读专业英文文献需要对相关领域具有一定的了解和掌握。要了解该领域的专业术语、概念和特点等，以便更好地理解文献内容。

（四）阅读技巧

阅读英文文献还需要掌握一些阅读技巧，比如建立阅读目标、识别关键信息、注重上下文等。可以通过阅读相关书籍和论文，积累阅读经验和技巧。

（五）练习阅读

最后，提高阅读能力还需要进行大量的实践和练习。可以选择一些专业英文文献，逐步提高难度，进行有针对性的阅读训练。

总之，提高专业英文文献阅读能力需要多方面的努力和学习。从扩充词汇量、学习句法结构、熟悉专业知识、掌握阅读技巧到大量练习，都是提高阅读能力的有效方法。

八、端正阅读专业英文文献的心态

阅读专业英文文献需要具备正确的心态，以下是一些建议：

（一）虚心学习

阅读专业英文文献不仅要有扎实的专业知识基础，还需要虚心学习，尊重不同领域、不同观点的知识和见解。

（二）科学思维

阅读专业英文文献需要具备科学思维，包括质疑、验证、探究等，需要通过理性思考来分析、评估和判断论文中的内容。

（三）深入思考

阅读专业英文文献需要深入思考，不能浅尝辄止，需要对文献中提出的问题进行更深入的思考和探讨。

（四）耐心坚持

阅读专业英文文献需要耐心坚持，可能会遇到一些难懂的术语和概念，需要耐心学习和积累，不能半途而废。

（五）合理分配时间

阅读专业英文文献需要合理分配时间，不能花费过多时间在某一个细节上，也不能匆匆忙忙地浏览全文，需要根据自己的需要和目的来确定时间分配。

（六）保持热情

阅读专业英文文献需要保持热情，对于自己感兴趣的领域和问题，需要保持积极向上的态度，不断深入学习和探索。

第二节　专业英文文献的翻译方法

一、概述

翻译专业英文文献需要遵循一定的方法和技巧。以下是它们的概述：

（一）熟悉原文

在进行翻译之前，必须先仔细阅读原文，并对其中的专业术语、概念和语法结构等有一个全面的理解。

(二) 确定翻译目标

在翻译之前，需要确定翻译的目标对象和领域，以便选用适当的翻译术语和表达方式。

(三) 选择合适的翻译工具

为了提高翻译效率和准确度，可以使用各种翻译软件和工具，如谷歌翻译、百度翻译、Translation Memory 等。

(四) 注重上下文

专业英文文献中常常存在复杂的句子结构和长难句，因此在进行翻译时，要注重上下文，理解句子的语境，以确保翻译的准确性和连贯性。

(五) 注意细节

在进行翻译时，需要注意一些细节问题，如标点符号、大小写、缩略词等，以确保翻译的规范性和精准度。

(六) 校对和修订

翻译完成后，需要进行校对和修订，以检查是否存在翻译错误和不合理之处，并进行相应的修改。

总之，翻译专业英文文献需要细心、耐心、严谨，并遵循一定的方法和技巧。要注重上下文、注意细节、选择合适的工具和方法，并进行校对和修订，以确保翻译的准确性和可靠性。

二、专业英文文献的翻译方法

专业英文文献的翻译方法包括逐句翻译法、整体理解翻译法和对比翻译法。

(一) 逐句翻译法

逐句翻译法是指将原文按照语序逐句进行翻译，重点考虑语言表达的准确性和精准度。该方法需要仔细理解每个单词和短语的含义，并根据上下文和语法结构进行准确翻译。

(二) 整体理解翻译法

整体理解翻译法是指先将原文整体理解，把握全文的主旨和大意，然后再进行翻译。该方法需要快速了解原文的结构和内容，确定其主题和目的，然后根据语言习惯和语境进行表达。

(三) 对比翻译法

对比翻译法是指将原文与译文对照，不断修正和完善翻译质量。该方法需要在逐句翻译或整体理解的基础上，与原文进行反复核对和比较，找出不准确或不恰当的翻译，进行修改和调整。

以上三种翻译方法各有优缺点，需要根据具体情况选择合适的方法。在实际应用时，也可以将它们结合起来，先用整体理解翻译法了解原文的主要内容和结构，再用逐句翻译法进行准确翻译，最后用对比翻译法进行修正和完善。

三、专业英文文献的翻译工具

包括机器翻译软件、在线翻译网站和翻译记忆软件。以下是它们的概述：

(一) 机器翻译软件

机器翻译软件是指利用计算机程序自动将一种语言的文本翻译成另一种语言的软件。目前市场上常见的机器翻译软件有 Google Translate、百度翻译、有道翻译等。虽然机器翻译软件的翻译速度快，但由于其翻译质量不稳定，需要人工校对。

(二) 在线翻译网站

在线翻译网站是指通过互联网提供翻译服务的网站，用户可以在网站上输入原文并选择翻译语种，即可获得翻译结果。目前比较常用的在线翻译网站有 DeepL、百度翻译、有道翻译等。虽然在线翻译网站的翻译质量相对机器翻译软件较好，但仍需要人工修改和校对。

(三) 翻译记忆软件

翻译记忆软件是指利用计算机程序存储先前已翻译过的文本和其翻译结果，并在后续翻译中根据相似度自动匹配现有文本，提高翻译速度和准确性的软件。常见的翻译记忆软件有 SDL Trados、MemoQ、Wordfast 等。翻译记忆软件的翻译质量相对较高且可以提高翻译效率，但需要事先建立翻译记忆库，并花费时间进行人工修订。

总之，以上三种专业英文文献翻译工具各有优缺点，需要根据实际需求和情况选择合适的工具。

四、提高专业英文文献翻译能力的方法

以下是提高专业英文文献翻译能力的几种方法：

（一）增强英语语言能力

提高英语语言能力是专业英文文献翻译的基础。可以通过听说读写等多种方式，提高语言水平。建议多看英文材料，了解英美环境和文化。

（二）学习专业知识

在翻译专业英文文献时，需要具有丰富的专业领域知识，才能更好地理解和翻译文献。因此，不仅需要掌握英语语言，还需要学习相关领域的专业术语、概念和特点等。

（三）熟悉翻译技巧

翻译并不是简单的语言转换，还需要运用一定的翻译技巧。例如，合理运用上下文信息、重点把握句子结构、准确翻译专业术语等。需要通过翻译练习和经验总结，积累翻译技巧。

（四）利用翻译工具

在翻译专业英文文献时，可以采用一些翻译工具，如机器翻译、翻译记忆和在线翻译网站等，以提高翻译效率。

（五）多练习

熟能生巧，在翻译专业英文文献方面也是如此。多进行实践、练习和尝试，不断积累经验，才能逐渐提高翻译水平。

总之，提高专业英文文献的翻译能力需要通过多种渠道学习和实践，不断积累经验和技巧。同时还需要注意保持语言的精准性和专业性，确保翻译质量。

第三节　农业科学领域专业词汇的构词法

一、概述

农业科学领域的专业英语论文中，常用的构词法包括前缀、后缀、复合词以及缩写词等。

（一）前缀

前缀：在单词词首加上一个字母或一组字母，可以改变词义或产生新词。例如：

（1）Unicellular（单细胞的）：uni-表示"单一的"，cellular 表示"由细胞组成的"。

（2）Heterozygous（杂合的）：hetero-表示"不同的"，zygous 表示"接合子的"。

（二）后缀

后缀：在单词词末加上一个字母或一组字母，可以改变词性、词义或产生新词。例如：

（1）Cytogenetics（细胞遗传学）：cyto-表示"细胞"，-genetics 表示"遗传学"。

（2）Monophyll（单叶）：mono-表示"单"，-phyll 表示"叶"。

（三）复合词

复合词：将两个或多个单词组合在一起形成新词。例如：

（1）Ecosystem（生态系统）：eco-来源于 ecology（生态学），system 表示"系统"。

（2）Photosynthesis（光合作用）：photo-表示"光的"，synthesis 表示"合成"。

（四）缩写词

缩写词：取单词的首字母组成的新词。例如：

（1）DNA（脱氧核糖核酸）：Deoxyribonucleic Acid。

（2）PCR（聚合酶链反应）：Polymerase Chain Reaction。

二、农业科学专业英语论文中常用的 150 个前缀及例词

农业科学领域专业词汇的前缀有很多，表 6.1 所列是常用的 150 个前缀及例词的含义。

表 6.1　农业科学领域常用的 150 个专业词汇前缀及例词的含义

前缀	意义	例词及其含义
a-	不；无	asexual（无性繁殖的）、amorphous（无定形的）
ab-	不；非；远离	abnormal（不正常的）、abiotic（非生物的）
ad-	向；增加	adapt（适应）、adherent（附着的）
aero-	空气；气体	aeroponics（气栽法）、aerobic（需氧的）
allo-	异；别	allopolyploid（异源多倍体）、allopatric（异地的）
ambi-	两个；含糊	ambipolar（两极的）、ambient（周围的）
amphi-	双重、两端	amphipathic（两性分子的）、amphimixis（混合受精）
ana-	分解；上升	analysis（分析）、anabolism（合成代谢）
ante-	在…之前；早期	antecedent（先驱的）、antecedent（前提）
antho-	花；花状物	anthocyanin（花青素）、anthotaxy（花序排列）
anti-	防止；对抗	antibiotic（抗生素）、antioxidant（抗氧化剂）
auto-	自动的；自己	autotroph（自养）、autophagy（自噬）
apo-	远离；离开	apoptosis（细胞凋亡）、apomixis（无融合生殖）
arch-	最好的；古代的	archaic（过时的）、architecture（结构）
auto-	自身的；自动的	autotroph（自养生物）、autogamous（自花授粉的）
bi-	两个；双重	binary（二进制的）、biennial（二年生的）
bio-	生命；生物	biology（生物学）、biota（生物群落）
cata-	向下；削弱	catalyst（催化剂）、catabolism（分解代谢）
cellul-	细胞	cellular（细胞的）、cellulose（纤维素）
cen-，centr-	中心；移动	centromere（着丝粒）、centrifugation（离心）
chemo-	化学；化学作用	chemotaxis（趋化性）、chemoreception（化感作用）
chlor-	绿色；氯	chlorophyll（叶绿素）、chlorine（氯）
chrom-，chromat-	彩虹；颜色	chromosome（染色体）、chromatography（色谱法）
chrono-	时间	chronobiology（生物节律学）、chronosequence（时间序列）

（续表）

前缀	意义	例词及其含义
circum-	围绕；周围	circumference（周长）、circumnutation（环旋运动）
co-	共同；一起	codominance（共显性）、coexist（共存）
con-	一起；共同	concentrate（浓缩）、conservation（保护）
contra-	相反的；对抗	contrast（反差）、contralateral（对侧）
corp-	组织；公司	corporation（公司）、corporate（企业的）
cryo-	极冷；低温	cryopreservation（低温保存）、cryotolerance（耐冷性）
crypt-	隐藏；秘密	cryptobiosis（休眠状态）、cryptochrome（色素蛋白质）
cyto-	细胞	cytoplasm（细胞质）、cytokinesis（胞质分裂）
de-	除去；相反	dehydrate（去水）、denature（使变性）
dec-	数量或程度的减少	decay（腐烂）、decimate（大批毁灭）
deca-	十；十倍的	decade（十年）、decaploid（十倍体）
def-，dep-	分开；离开	defoliation（落叶）、depart（离开）
di-	双；两种	diploid（二倍体的）、disaccharide（二糖）
dia-	通过；穿过	diagnose（诊断）、diameter（直径）
dis-	不；分开	distribution（分布）、dissolution（溶解）
eco-	环境	ecology（生态学）、ecosystem（生态系统）
ecto-	外部；表面上	ectoderm（外胚层）、ectotrophic（体外营养的）
em-	在…起作用；使…	emergence（出苗）、emasculation（去雄）
embry-	胚芽；胚胎	embryo（胚）、embryonic（萌芽期的）
en-	使…；在…之内	enhancer（增强子）、enrichment（富集）
endo-	内部；内膜上	endocytosis（内吞）、endosperm（胚乳）
ento-	昆虫	entomology（昆虫学）、entomophilous（昆虫授粉的）
epi-	在…之上；超过	epidermis（表皮层）、epigenetics（表观遗传学）
eu-	真正的；好的；完善的	eukaryote（真核生物）、eutrophication（富营养化）
ex-	向外；除去	exclude（排除）、expression（表达）
exo-	外部；外面	exocytosis（胞吐）、exon（外显子）
extra-	在…之外；超出	extracellular（细胞外的）、extractive（提取的）
fer-	带来；生育	fertile（肥沃的）、fertilization（受精）
ferro-	铁	ferrous（亚铁的）、ferrosols（富铁土）
fi-，fy-	发生；调节	fiber（纤维）、fytochemical（植物化学物质）

(续表)

前缀	意义	例词及其含义
fil-	纤维；线	filament（花丝）、filamentous（纤维状的）
flagell-	鞭毛；鞭形体	flagellum（鞭毛）、flagellar（鞭毛的）
foli-	叶子	foliage（叶片）、foliar（叶片的）
fore-	在前面的	foreground（前景）、forecast（预测）
fos-	化合物或元素的来源	fosssil（化石的）、foster（培养）
fract-	断裂；破碎	fracture（断裂）、fractal（分形）
fruct-	果糖；含果糖的	fructose（果糖）、fructan（果聚糖）
fungi-	真菌	fungicide（杀真菌剂）、fungistasis（抑菌状态）
gen-，geno-	产生；起源	gene（基因）、genotype（基因型）
geo-	地球；地理学	geography（地理学）、geotropism（向地性）
germ-	胚芽；种子	germplasm（种质）、germination（发芽）
giga-	十亿；庞大的	gigabase（千兆碱基）、gigantic（巨大的）
glyco-	糖；含糖的	glycolysis（糖解作用）、glycosylation（糖基化）
gram-	写，描绘；重量计	grammar（语法）、gramineae（禾本科）
graph-	描述；记录；画	graphic（图形化）、graphical（图形的）
hemi-	半	hemiparasite（半寄生）、hemizygous（半等位基因）
hetero-	不同的；异质的	heterozygous（杂合的）、heterotrophic（异养的）
hist-	组织	histone（组蛋白）、histidine（组氨酸）
homo-	相同的；同种的	homogenize（均质化）、homogeneous（同质的）
hormo-	激素	hormone（激素）、hormonal（激素的）
hydr-，hydro-	水；水合物	hydroponics（水培法）、hydrophobic（疏水的）、hydraulic（水力的）
hyper-	过量的；更高的	hyperploid（多倍体）、hypertonic（高渗的）
hypo-	不足的；低于	hypocotyl（下胚轴）、hypogeic（地下的）
ideo-	思想；观念	ideotype（观念型）、ideogram（字符造型）
idio-	个人的；特殊的	idiopathic（特发性）、idioblast（异型细胞）
immun-	免疫；抵抗力	immunology（免疫学）、immunogenetics（免疫遗传学）
in-	不；无	inactive（不活跃）、inhibition（抑制）
infra-	基础；下方	infrared（红外线的）、infraspecific（子种级别）
inter-	在…之间	intercellular（细胞间的）、intercrop（间作）
intra-	在…之内	intracellular（细胞内的）、intraseasonal（季节内）

（续表）

前缀	意义	例词及其含义
iso-	等同的；同种的	isometric（等距的）、isogenic（同系）
kary-	核；细胞核	karyotype（染色体组型）、karyogamy（核融合）
ker-	角质；硬化	keratin（角质素）、kerogen（干酪根），一种沉积有机质
kin-，kine-	激素；中心	kinase（激酶）、kinetochore（着丝点）
kilo-	千；一千	kilobase（千碱基对）、kilogram（千克）
lacto-	乳；含乳的	lactose（乳糖）、lactobacillus（乳酸菌）
lexi-	语言；文字	lexical（词汇的）、lexicon（辞典）
lip-	脂肪	lipase（脂肪酶）、lipid（脂质）
macro-	大型的；大量的	macromolecule（大分子）、macronutrient（大量营养素）
mal-	坏的；疾病	maladaptation（不适应）、malfunction（机能障碍）
medi-	中间；调节	medial（中央的）、median（中位数）
mega-	百万；巨大的	megabase（兆碱基）、megaspore（大孢子）
mer-	部分；分裂	meristem（分生组织）、merogony（部分有性生殖）
meso-	中等的；中央的	mesophyll（叶肉组织）、mesoderm（中胚层）
meta-	超越；转化	metapopulation（亚种群）、metabolism（新陈代谢）
micro-	微小的；百万分之一	microscope（显微镜）、microorganism（微生物）
milli-	千分之一；千分	milligram（毫克）、milliliter（毫升）
mis-	错误的；不当的	misclassification（分类错误）、mismatch（不匹配）
mit-	送出；分裂	mitochondrion（线粒体）、mitotic（有丝分裂的）
mono-	单一的；单个的	monoculture（单作）、monosaccharide（单糖）
morph-	形态；变形	morphology（形态学）、morphogenesis（形态发生）
multi-	许多的；多重的	multiple（多重的）、multicellular（多细胞的）
mut-	突变；改变	mutation（突变）、mutagenic（致突变的）
myco-	真菌	mycology（真菌学）、mycorrhiza（菌根）
nano-	十亿分之一；纳米	nanometer（纳米）、nanotechnology（纳米技术）
necro-，necros-	坏死；腐烂	necrosis（坏死）、necrotroph（腐生菌）
neo-	新的；现代的	neoplasia（新生的组织）、neo-Darwinism（新达尔文主义）
non-	不；非	nonrenewable（不可再生的）、non-enzymatic（非酶促的）
ob-	在…之下；遮盖	observation（观察）、obstruct（阻碍）
oc-	隐藏；遮盖	occlude（闭塞）、occupy（占据）

（续表）

前缀	意义	例词及其含义
oligo-	少量；少数的	oligomer（低聚物）、oligosaccharide（低聚糖）
omni-	全部的；所有的	omnifarious（多种的）、omnipotent（万能的）
op-	对；面向	opposite（相反的）、operon（操纵子）
osmo-	渗透；渗透压	osmoregulation（渗透调节）、osmolarity（渗透压）
out-	外部的；离开	outbreed（远缘繁殖）、outlook（前景）
para-	旁边的；类似的	parasite（寄生物）、parameter（参数）
peri-	环绕；周围的	pericarp（果皮）、pericycle（中柱鞘）
photo-	光；光合作用	photosynthesis（光合作用）、photoperiodism（光周期）
phyto-	植物；植物学	phytochrome（植物修复）、phytohormone（植物激素）
plasm-	细胞质；原生质体	plasmid（质粒）、plasmodesmata（胞间连丝）
poly-	多；多种的	polymer（聚合物）、polyculture（混种）
post-	在…之后	postharvest（采后）、posttranscriptional（转录后）
pre-	在…之前；预先	preharvest（采前）、precede（先于）
prim-	第一；最初的	primary（最初的、主要的）、primer（引物）
pro-	促进；赞成	promote（促进）、promoter（启动子）
proto-	原始；原形	protoderm（初生表皮）、protoplast（原生质体）
pseudo-	假；伪；虚	pseudoscience（伪科学）、pseudofruit（虚果）
retro-	向后；倒退	retrograde（后退的）、retrotransposon（反转座子）
rhizo-	根；根状物	rhizosphere（根际）、rhizome（根状茎）
sap-	液汁；汁液	sapwood（边材）、saprophyte（腐生植物）
self-	自我；自己的	self-pollination（自交）、self-incompatibility（自交不亲和）
semi-	半；不完全的	semipermeable（半透过性的）、semiarid（半干旱的）
sub-	在…之下；次要的	subcellular（亚细胞的）、substance（物质）
super-	超过；超级的	superoxide（超氧化物）、superfamily（超家族）
supra-	在…之上；超越	supramolecular（超分子的）、suprathermal（超热的）
syn-	共同；一起	synthesis（合成）、synergistic（协作的）
tele-	远程的；远距离的	telemetry（遥测）、teleoperation（远程操作）
therm-，thermo-	热；温度	thermal（热的）、thermography（热像技术）
trans-	超过；横越	transport（运输）、transcriptional（转录）
tri-	三；三个	trisomy（染色体三体性）、tripeptide（三肽）

（续表）

前缀	意义	例词及其含义
ultra-	超过；超级的	ultrastructure（超微结构）、ultrasonic（超声的）
uni-	单一的；一个	uniform（一致性）、unisexual（单性的）
ur-	尿；尿中的	urea（尿素）、uracil（尿嘧啶）
vascul-	维管的；脉管的	vascular（管状的）、vasculature（维管系统）
xeno-	异种的；异国的	xenogamy（异交）、xenograft（异种嫁接）

三、农业科学专业英语论文中常用的 100 个后缀及例词

农业科学领域专业词汇的后缀有很多，表 6.2 所列是常用的 100 个后缀及例词的含义。

表 6.2　农业科学领域常用的 100 个专业词汇后缀及例词的含义

后缀	含义	示例
-able	能够；可以	adaptable（适应性强的）、sustainable（可持续的）
-aceae	科	Poaceae（禾本科植物）、Rosaceae（蔷薇科）
-aceous	像…的	herbaceous（草本植物的）、tuberaceous（块茎的）
-acity	质量；状态	capacity（容量）、tenacity（韧性）
-ad	向；附着或方向	palisad（栅状组织）、centrad（中心方向的）
-ae, -i	复数形式	algae（藻类）、fungi（真菌）
-al	形容词性后缀，表示…的	annual（一年生的）、basal（基部的）
-amine	胺；氨基化合物	histamine（组胺）、tyramine（酪胺）
-ance, -ancy	名词性后缀，表示…的状态或品质	resistance（抗性）、dormancy（休眠）
-ant	形容词性后缀，表示…的	dormant（休眠的）、mutant（突变体）
-anthus	花	Lisianthus（洋桔梗）、helianthus（向日葵花）
-ar	形容词性后缀，表示…的	cellular（细胞的）、molecular（分子的）
-ard	人、物及其他	vanguard（先锋）、standard（标准）
-aria	地点；场所	arboretaria（植物园）、aquaria（水族馆）
-arium	存放物	herbarium（植物标本馆）、terrarium（生物育养箱）

（续表）

后缀	含义	示例
-ase	酶	amylase（淀粉酶）、protease（蛋白酶）
-ate	动词性后缀，表示…化	pollinate（传粉的）、germinate（发芽的）
-ation	名词性后缀，表示…过程或状态	germination（萌发）、domestication（驯化）
-blast	发生的；原基	genoblast（成熟生殖细胞）、endoblast（内胚层）
-biosis	生存；生态	symbiosis（共生）、aerobiosis（需氧生存）
-calyx	萼片	epicalyx（副萼）、macrocalyx（大萼）
-carp	果实	pericarp（果皮）、endocarp（内果皮）
-carpous	子房	syncarpous（合心子房）、apocarpous（独心子房）
-cellulose	纤维素	hydrocellulose（水解纤维素）、nanocellulose（纳米纤维素）
-cide	杀死	herbicide（除草剂）、fungicide（杀真菌剂）
-ciduous	自然脱落的	deciduous（落叶性的）、semidecious（半落叶的）
-cola	与环境、病害相关的	corticola（树皮生的）、graniticola（岩生生物）
-colous	生活在某种环境中的	terricolous（土长的）、aquicolous（水生的）
-coma	伴随状态	Trichocoma（须刷菌）、mycoma（真菌营养体）
-coccus	球菌	staphylococcus（葡萄球菌）、micrococcus（小球菌属）
-crine	分泌	exocrine（外分泌的）、endocrine（内分泌的）
-cula	小；株	vesicula（小囊）、radicula（胚根）
-cule	微小；小分子	molecule（分子）、canalicule（毛细管道）
-culture	名词性后缀，表示…的培养	monoculture（单作）、polyculture（混种）
-dom	状态；领域	kingdom（界）、random（随机）
-eae	科名后缀	Poaceae（禾本科）、Malvaceae（锦葵科）
-el	小；株	kernel（谷粒）、pedicel（花柄）
-ella	小；微小	lenticella（皮孔）、proteinella（蛋白体）
-ellus	小；微小	flavellus（黄细胞）、ocellus（油胞）
-en	变得；使…化为	lengthen（延长）、ripen（成熟）
-ence	表示状态、性质或过程	dependence（依赖）、resilience（弹性）
-ene	烯类化合物	ethylene（乙烯）、carotene（胡萝卜素）
-ensis	地点归属	sinensis（中国的）、australensis（澳洲的）
-ent	形容词性后缀，表示…的	quiescent（静止的）、nutrient（养分的）

（续表）

后缀	含义	示例
-er	名词性后缀，表示做…的人或机械	researcher（研究人员）、weeder（除草机）
-ery	表示特定性质、状态或场所	vinery（葡萄园）、nursery（苗圃）
-escent	变化；时期	efflorescent（盛花期）、florescent（开花期）
-esis	过程；状态	biosynthesis（生物合成）、photosynthesis（光合作用）
-form	表示形状、形式	multiform（多样化的）、cupuliform（杯状的）
-gen	名词性后缀，表示…的产生或形成因素	mutagen（诱变剂）、pathogen（病原体）
-genesis	起源；生成	endogenesis（内生作用）、ontogenesis（发育过程）
-genus	属	Zeagenus（玉米属）、monogenus（单属）
-germ	与种子、胚、生殖相关	microgerm（微生物）、germ（繁殖）
-grade	分类；等级	infraspecificgrade（品种以下分类）、first-grade（一等品）
-gram	名词性后缀，表示…的图表或克	histogram（直方图）、microgram（微克）
-gynous	雌蕊的	androgynous（雄雌同体的）、heterostylygynous（异位异型花的）
-ia	状态、疾病或表现	bacteria（细菌）、pathogenia（致病性）
-ic	形容词性后缀，表示…的	photosynthetic（光合作用的）、botanic（植物学的）
-ica	类，表示…的	agronomica（农艺学的）、botanica（植物园）
-id	像…的	acid（酸性的）、arid（干旱的）
-ide	化合物	pesticide（杀虫剂）、carbide（碳化物）
-ification	名词后缀，表示使…化	classification（分类）、verification（验证）
-ify	动词后缀，变成…的过程	calcify（钙化）、diversify（多样化）
-ii, -ius	具有…特征的	infectius（传染的）、pollinarius（粉状的）
-in	蛋白质	glutenin（麦谷蛋白）、actin（肌动蛋白）
-ine	表示化合物、碱、药物等的名称及其含义	caffeine（咖啡因）、saccharine（糖精）
-ious	充分的	copious（丰富的）、contagious（传染性的）
-ism	繁殖；过程	organism（生物体）、physiologism（生理现象）
-ium	物质名；地点名	magnesium（镁）、herbarium（植物标本馆）
-ive	表示有倾向性或有关的	reproductive（生殖的）、photoreceptive（感光的）

（续表）

后缀	含义	示例
-ize，-ise	动词性后缀，表示使…化	fertilize（施肥）、utilize（利用）
-karyon	细胞核	eukaryon（真核生物）、plasmakaryon（原生质体）
-le	…的小物体	molecule（分子）、vacuole（液泡）
-let	小；株	pollenlet（花粉）、leaflet（小叶）
-ling	小的；年幼的	seedling（幼苗）、sapling（小树苗）
-logist	名词性后缀，表示从事某一领域的人	biologist（生物学家）、botanologist（植物学家）
-logy	名词性后缀，表示…学	biology（生物学）、ecology（生态学）
-lysis	分解；溶解；消退	hydrolysis（水解作用）、autolysis（自溶作用）
-mata	复数	chromosomata（染色体群）、stigmata（气孔，红斑）
-ment	名词性后缀，表示…的过程或结果	development（发育）、improvement（改良）
-merous	部分数	monomerous（单数的）、schizomerous（裂成部分的）
-metry	测量	biometry（生物计量学）、spectrometry（光谱测量）
-morph	形态	polymorph（多态型）、xenomorph（异形态）
-myces	菌类	actinomyces（放线菌）、streptomyces（链霉菌）
-nomy	学科	taxonomy（分类学）、agronomy（农学）
-oid	相似、类似之物	amyloid（淀粉状的）、carotenoid（类胡萝卜素）
-ose	糖	fructose（果糖）、glucose（葡萄糖）
-osis	表示状态；发展情况	necrosis（坏死）、symbiosis（共生）
-phyll	叶子	mesophyll（叶肉组织）、chlorophyll（叶绿素）
-ploid	染色体倍数	triploid（三倍体）、haploid（单倍体）
-ptosis	落下	apoptosis（细胞凋亡）、pyroptosis（细胞焦亡）
-phyte	具有某种特征之植物	xerophyte（旱生植物）、halophyte（盐生植物）
-plasm	组织或细胞的浆液	cytoplasm（细胞质）、protoplasm（原生质）
-some	体、体细胞或有形成分	ribosome（核糖体）、lysosome（溶酶体）
-spore	孢子	ascospore（子囊孢子）、conidiospore（分生孢子）
-taxis	有序运动；向性运动	phototaxis（趋光性运动）、chromotaxis（色素趋向性运动）
-theca，-thecium	囊、袋或盒子	sporotheca（孢子囊）、ascocarpthecium（子囊器）

（续表）

后缀	含义	示例
-trophy	营养	autotrophy（自养）、Mycotrophy（菌根营养）
-ule	小型；小的	ovule（胚珠）、nodule（结节）
-y	表示状态或质量	cavity（空腔）、density（密度）

四、农业科学专业英语论文中常用的 150 个复合词

农业科学领域的复合词有很多，表 6.3 所列是常用的 150 个复合词及其含义。

表 6.3　农业科学领域常用的 150 个复合词及其含义

单词	含义	复合词示例
abiotic	非生物的	abiotic stress（非生物胁迫）
abscisic	脱落的	abscisic acid（ABA）（脱落酸）
aerobic	好氧的	aerobic respiration（有氧呼吸）
aging	老化	accelerated aging（加速老化）
agriculture	农业	precision agriculture（精准农业）
agronomy	农学	crop agronomy（作物农学）
agroforestry	农林耕种	agroforestry system（农林耕种系统）
allelic	等位的	allelic diversity（等位基因多样性）
allelopathy	化感作用	allelopathy effect（化感作用效应）
anthracnose	炭疽病	anthracnose resistance（炭疽病抗性）
apomixis	无性生殖	apomixis reproduction（无性生殖繁殖）
architecture	构型	plant architecture（株型）
arid	干旱的	arid zone agriculture（干旱地区农业）
breeding	育种	crop breeding（作物育种）
biological	生物的	biological control（生物防治）
biotechnology	生物技术	plant biotechnology（植物生物技术）
botany	植物学	economic botany（经济植物学）
calcium	钙	calcium carbonate（碳酸钙）
canopy	冠层	crop canopy（作物冠层）

(续表)

单词	含义	复合词示例
carbon	碳	carbon footprint（碳足迹）
cell	细胞	cell culture（细胞培养）
cellulose	纤维素	cellulose content（纤维素含量）
cereal	谷类	cereal crops（谷类作物）
chlorophyll	叶绿素	chlorophyll fluorescence（叶绿素荧光）
chromosome	染色体	chromosome number（染色体数目）
clone	克隆	plant clone（植物克隆）
cold	寒冷	cold tolerance（耐冷性）
conservation	保育	crop conservation（作物保育）
crop	作物	cash crop（经济作物）
cultivar	品种	cultivar selection（品种选择）
cultivation	栽培	crop cultivation（作物栽培）
cytology	细胞学	plant cytology（植物细胞学）
defoliation	落叶	chemical defoliation（化学落叶）
dehydration	脱水	plant dehydration（植物脱水）
disease	病害	disease incidence（发病率）
dormancy	休眠	seed dormancy（种子休眠）
drought	干旱	drought tolerance（耐旱性）
dwarfing	矮化	dwarfing gene（矮化基因）
ecology	生态学	plant ecology（植物生态学）
embryo	胚芽	plant embryo（植物胚芽）
endophyte	内生菌	grass endophyte（草地内生菌）
entomology	昆虫学	insect entomology（昆虫学）
environmental	环境的	environmental stress（环境胁迫）
enzyme	酶	enzyme activity（酶活性）
epigenetic	表观遗传学	epigenetic regulation（表观遗传调控）
erosion	侵蚀	soil erosion（土壤侵蚀）
farming	农业	eco-friendly farming（生态友好型农业）
fatty	脂肪的	fatty acid（脂肪酸）
fertility	肥力	soil fertility（土壤肥力）

（续表）

单词	含义	复合词示例
fertilizer	肥料	slow-release fertilizer（缓释肥料）
fibre	纤维	fibre crop（纤维作物）
field	农田	field management（农田管理）
flooding	洪涝	flooding tolerance（耐涝性）
floret	小花	wheat floret（小麦小花）
flowering	开花	photoperiodic flowering（光周期性开花）
foliar	叶面的	foliar nutrition（叶面施肥）
food	食品	food security（食品安全）
fruiting	结实	fruiting body（结实体）
fungal	真菌性的	fungal disease（真菌性病害）
gene	基因	gene pool（基因库）
genetic	遗传的	genetic diversity（遗传多样性）
genotype	基因型	plant genotype（植物基因型）
germination	萌发	seed germination（种子萌发）
germplasm	种质	germplasm conservation（种质资源保护）
girdling	环剥	tree girdling（树皮环剥）
grafting	嫁接	plant grafting（植物嫁接）
grain	谷物	grain filling（结实期）
green house	温室	greenhouse gas（温室气体）
growth	生长	plant growth and development（植物生长发育）
harvest	收获	harvest index（收获指数）
heat	热的	heat tolerance（耐热性）
herbicide	除草剂	selective herbicide（选择性除草剂）
hormone	激素	hormone signaling（激素信号）
host	寄主	host plant（寄主植物）
hybrid	杂交	plant hybrid（植物杂交）
immunology	免疫学	plant immunology（植物免疫学）
insecticide	杀虫剂	systemic insecticide（内吸性杀虫剂）
inter-cropping	间作	inter-cropping system（间作系统）
intracellular	细胞内的	intracellular protein（细胞内蛋白质）

(续表)

单词	含义	复合词示例
in vitro	体外	in vitro culture（体外培养）
irrigation	灌溉	drip irrigation（滴灌）
land	土地	land degradation（土地退化）
leaf	叶子	plant leaf（植物叶子）
legume	豆类	legume crops（豆类作物）
light	光线	plant light response（植物光响应）
lignin	木质素	lignin content（木质素含量）
lipid	脂质	lipid metabolism（脂质代谢）
marker	标记	marker-assisted selection（MAS）（标记辅助选择）
mechanized	机械化的	mechanized farming（机械化农业）
microbial	微生物的	microbial diversity（微生物多样性）
microbiology	微生物学	soil microbiology（土壤微生物学）
mitotic	有丝分裂	mitotic cell division（有丝分裂细胞分裂）
molecular	分子的	molecular marker（分子标记）
mulching	覆盖	mulching technique（覆盖技术）
mycorrhiza	菌根	arbuscular mycorrhiza（丛枝菌根）
nematode	线虫	plant nematode（植物线虫）
nitrogen	氮	nitrogen fixation（固氮作用）
nutrient	营养的	nutrient uptake（营养吸收）
organic	有机的	organic matter（有机物）
organism	生物	genetically modified organism（转基因生物）
pest	害虫	pest control（防治害虫）
pesticide	杀虫剂	pesticide application（农药施用）
photosynthesis	光合作用	photosynthesis rate（光合速率）
physiology	生理学	plant physiology（植物生理学）
phenology	物候	crop phenology（农作物物候）
phenotypic	表型的	phenotypic trait（表型性状）
phloem	韧皮部	phloem transport（韧皮部输送）
phylogeny	系统发育	molecular phylogeny（分子系统发育）
phytopathology	植物病理学	phytopathology research（植物病理学研究）

（续表）

单词	含义	复合词示例
physiology	生理学	plant physiology（植物生理学）
pigment	色素	photosynthetic pigment（光合色素）
pollination	授粉	insect pollination（昆虫授粉）
post-harvest	采后	post-harvest technology（采后技术）
precision	精准	precision nutrient management（精准养分管理）
productivity	生产力	productivity improvement（生产力提高）
protection	保护	plant protection（植物保护）
protein	蛋白质	plant protein content（植物蛋白质含量）
quality	品质	plant quality traits（植物品质性状）
radiation	辐射	plant radiation response（植物辐射响应）
regulator	调节剂	growth regulator（生长调节剂）
resistance	抗性	disease resistance（抗病性）
root	根	root system（根系）
rotation	轮作	crop rotation（轮作）
rust	锈病	wheat rust（小麦锈病）
salinity	盐分	salinity stress（盐胁迫）
seed	种子	seed production（种子生产）
seedling	实生苗	seedling emergence（出苗）
senescence	衰老	plant senescence（植物衰老）
significant	重要的	significant traits（重要性状）
soil	土壤	plant-soil interactions（植物-土壤相互作用）
stage	时期	growth stage（生长期）
stress	压力	drought stress（干旱压力）
sustainable	可持续的	sustainable agriculture（可持续农业）
symbiosis	共生关系	plant-microbe symbiosis（植物-微生物共生关系）
synthetic	合成的	synthetic seed（合成种子）
taxonomy	分类学	plant taxonomy（植物分类学）
temperature	温度	temperature stress（温度胁迫）
tillage	耕作	tillage practices（耕作制度）
tissue	组织	plant tissue culture（植物组织培养）

（续表）

单词	含义	复合词示例
trait	性状	trait improvement（性状改良）
transgenic	转基因的	transgenic technology（转基因技术）
transpiration	蒸腾作用	transpiration rate（蒸腾速率）
variety	品种	variety evaluation（品种评价）
vascular	维管的	vascular bundle（维管束）
vegetative	营养的	vegetative propagation（营养繁殖）
virus	病毒	virus diseases（病毒病）
water	水	water potential（水势）
weed	杂草	weed control（杂草控制）
wilt	枯萎的	wilt disease（枯萎病）
yield	产量	yield potential（产量潜力）

五、农业科学专业英语论文中常用的 100 个缩写词

农业科学领域的缩写词有很多，表 6.4 所列是常用的 100 个缩写词及其含义。

表 6.4　农业科学领域常用的 100 个缩写词及其含义

缩写词	全称	中文释义
ABA	Abscisic acid	脱落酸
AFLP	Amplified fragment length polymorphism	扩增片段长度多态性
ANOVA	Analysis of variance	方差分析
APX	Ascorbate peroxidase	抗坏血酸过氧化酶
ARF	Auxin response factor	生长素响应因子
ATP	Adenosine triphosphate	腺苷三磷酸
AUX	Auxin	生长素
BLAST	Basic local alignment search tool	基本局部序列比对工具
bp	Base pair	碱基对
BRs	Brassinosteroids	油菜甾醇类化合物
BSA	Bovine serum albumin	牛血清白蛋白

（续表）

缩写词	全称	中文释义
Bt	*Bacillus thuringiensis*	苏云金杆菌
C/N	Carbon to nitrogen ratio	碳氮比
CaM	Calmodulin	钙调蛋白
CAT	Catalase	过氧化氢酶
cDNA	Complementary DNA	互补 DNA
CDPK	Calcium-dependent protein kinase	钙依赖性蛋白激酶
circRNA	circular RNA	环状 RNA
CKs	Cytokinins	细胞分裂素
cM	Centimorgan	厘摩（遗传图谱距离单位）
CoA	Coenzyme A	辅酶 A
CRISPR	Clustered regularly interspaced short palindromic repeats	集束规律间隔短回文重复序列
CTAB	Cetyltrimethylammonium bromide	溴代十六烷基三甲基铵
CV	Coefficient of variation	变异系数
cv.	Cultivar	栽培品种
DH	Doubled haploid	加倍单倍体
DIGE	Difference gel electrophoresis	差异凝胶电泳
DNA	Deoxyribonucleic acid	脱氧核糖核酸
dsDNA	double-stranded DNA	双链 DNA
DUS	Distinctness, uniformity and stability	特异性、一致性和稳定性
EC	Electrical conductivity	电导率
EDTA	Ethylenediaminetetraacetic acid	乙二胺四乙酸
ELISA	Enzyme-linked immunosorbent assay	酶联免疫吸附测定法
ESI-MS	Electrospray ionization mass spectrometry	电喷雾离子化质谱
EST	Expressed sequence tags	外显子单拷贝序列标签
ET	Ethylene	乙烯
ETC	Electron transport chain	电子传递链
F_1	First filial generation	第一代杂交后代
FAD	Flavin adenine dinucleotide	黄素腺嘌呤二核苷酸
FISH	Fluorescence in situ hybridization	原位荧光杂交

（续表）

缩写词	全称	中文释义
Fv/Fm	Maximum quantum yield of photosystem II	光系统II最大量子产率
GA	Gibberellins	赤霉素
GC	Gas chromatography	气相色谱
GO	Gene ontology	基因本体论
GM	Genetically modified	转基因
GMO	Genetically modified organism	转基因生物
GST	Gene sequence tag	基因序列标签
GUS	Glucuronidase	β-葡萄糖醛酸酶
GWAS	Genome-wide association study	全基因组关联研究
HPLC	High-performance liquid chromatography	高效液相色谱
IAA	Indole-3-acetic acid	吲哚乙酸
IBA	Indole-3-butyric acid	吲哚-3-丁酸
ICP	Inductively coupled plasma	电感耦合等离子体
InDel	Insertion-deletion polymorphism	插入-缺失多态性
ISSR	Inter-simple sequence repeat	简单序列内重复
JA	Jasmonic acid	茉莉酸
Km	Michaelis constant	米氏常数
LC	Liquid chromatography	液相色谱
LD	Linkage disequilibrium	连锁不平衡
lncRNA	long non-coding RNA	长链非编码 RNA
MAB	Marker-assisted breeding	标记辅助育种
MAS	Marker-assisted selection	标记辅助选择
MDA	Malondialdehyde	丙二醛
miRNA	micro RNA	微小 RNA
mRNA	messenger RNA	信使核糖核酸
MS	Murashige and skoog	MS（培养基）
MS/MS	Tandem mass spectrometry	串联质谱
mtDNA	mitochondrial DNA	线粒体 DNA
NAA	Naphthaleneacetic acid	萘乙酸
NAD	Nicotinamide adenine dinucleotide	烟酰胺腺嘌呤二核苷酸

(续表)

缩写词	全称	中文释义
PCR	Polymerase chain reaction	聚合酶链反应
PEG	Polyethylene glycol	聚乙二醇
pH	Potential hydrogen	酸碱度
PTGS	Post-transcriptional gene silencing	转录后基因沉默
qPCR	Quantitative PCR	定量 PCR
QTL	Quantitative trait loci	定量性状位点
RAPD	Random amplified polymorphic DNA	随机扩增 DNA 多态性
RFLP	Restriction fragment length polymorphism	限制性片段长度
RNA	Ribonucleic acid	核糖核酸
RNAi	RNA interference	RNA 干扰
rRNA	ribosomal RNA	核糖体 RNA
ROS	Reactive oxygen species	活性氧物质
RT-PCR	Reverse transcription PCR	逆转录 PCR
SA	Salicylic acid	水杨酸
SCAR	Sequence characterized amplified region	序列特异性扩增区
SD	Standard deviation	标准差
SDS-PAGE	Sodium dodecyl sulfate-polyacrylamide gel electrophoresis	十二烷基硫酸钠-聚丙烯酰胺凝胶电泳
siRNA	small interfering RNA	小干扰 RNA
snoRNA	small nucleolar RNA	小核仁 RNA
SNP	Single nucleotide polymorphism	单核苷酸多态性
snRNA	small nuclear RNA	小核 RNA
SOD	Superoxide dismutase	超氧化物歧化酶
ssDNA	Single-stranded DNA	单链 DNA
SSR	Simple sequence repeat	简单序列重复
STR	Short tandem repeat	短串联重复序列
tRNA	transfer RNA	转运 RNA
UV	Ultra-violet	紫外线
VIGS	Virus-induced gene silencing	病毒介导的基因沉默
WUE	Water use efficiency	水分利用效率
Ψw	Water potential	水势

参考文献

姜巨全. 2011. 生物专业英语 [M]. 北京：化学工业出版社.

蒋悟生. 2010. 生物专业英语 [M]. 北京：高等教育出版社.

苗艳芳，李友军. 2007. 农科专业英语 [M]. 北京：气象出版社.

谭万忠，王进军. 2015. 农业与生物科学专业英语 [M]. 北京：科学出版社.

赵毓琴，常梅. 2019. 英语科技文献阅读 [M]. 北京：外语教学与研究出版社.

Ademe M S, He S P, Pan Z E, et al. 2017. Association mapping analysis of fiber yield and quality traits in Upland cotton (*Gossypium hirsutum* L.). Molecular Genetics and Genomics, 292 (6): 1267-1280.

Geng X L, Sun G F, Qu Y J, et al. 2020. Genome-wide dissection of hybridization for fiber quality- and yield-related traits in upland cotton. The Plant Journal, 104 (5): 1285-1300.

Sarfraz Z, Iqbal M S, Geng X L, et al. 2021. GWAS mediated elucidation of heterosis for metric traits in cotton (*Gossypium hirsutum* L.) across multiple environments. Frontiers in Plant Science, 12: 565552.

Sarfraz Z, Iqbal M S, Pan Z E, et al. 2018. Integration of conventional and advanced molecular tools to track footprints of heterosis in cotton. BMC Genomics, 19 (1): 776.

Sun R R, Li C Q, Zhang J B, et al. 2017. Differential expression of microRNAs during fiber development between fuzzless-lintless mutant and its wild-type allotetraploid cotton. Scientific Reports, 7 (1): 3-10.

Zhang X, Feng Y, Khan A, et al. 2022. Quantitative proteomics-based analysis reveals molecular mechanisms of chilling tolerance in grafted cotton seedlings. Agronomy, 12 (5): 1152.

Zhang X, Khan A, Zhou R Y, et al. 2022. Grafting in cotton: A mechanistic approach for stress tolerance and sustainable development. Industrial Crops and Products, 175: 114227.

Zhang X, Kong X J, Zhou R Y, et al. 2020. Harnessing perennial and indeterminant growth habits for ratoon cotton (*Gossypium* spp.) cropping. Ecosystem Health and Sustainability, 6: 1715264.

Zhang X, Li C Q, Wang X Y, et al. 2012. Genetic analysis of cryotolerance in cotton during the overwintering period using mixed model of major gene and polygene. Journal of Integrative Agriculture, 11 (4): 537-544.

Zhang X, Wang G, Xue H Y, et al. 2020. Metabolite profile of xylem sap in cotton seedlings is changed by K deficiency. Frontiers in Plant Science, 11: 592591.

Zhang X, Xue H Y, Khan A, et al. 2021. Physio-biochemical and proteomic mechanisms of coronatine induced potassium stress tolerance in xylem sap of cotton. Industrial Crops and Products, 173: 114094.

Zhang X, Yang Q, Zhou R Y, et al. 2022. Perennial cotton ratoon cultivation: A sustainable method for cotton production and breeding. Frontiers in Plant Science, 13: 882610.

Zhang X, Zhang Z Y, Wang Q L, et al. 2013. Effects of rootstocks on cryotolerance and overwintering survivorship of genic male sterile lines in upland cotton (*Gossypium hirsutum* L.). Plos One, 8 (5): e63534.

Zhang X, Zhang Z Y, Zhou R Y, et al. 2020. Ratooning Annual Cotton (*Gossypium* spp.) for Perennial Utilization of Heterosis. Frontiers in Plant Science, 11: 554970.

Zhang Z Y, Xin W W, Wang S F, et al. 2015. Xylem sap in cotton contains proteins that contribute to environmental stress response and cell wall development. Functional & Integrative Genomics, 15 (1): 17-26.

Zhang Z Y, Zhang X, Hu Z B, et al. 2015. Lack of K-dependent oxidative stress in cotton roots following coronatine-induced ROS accumulation. Plos One, 10 (5): e0126476.

Zhang Z Y, Zhang X, Wang S F, et al. 2013. Effect of mechanical stress on cotton growth and development. Plos One, 8 (12): e82256.